MATERIALS SCIENCE AND TECH

ADHESIVES

MECHANICAL PROPERTIES, TECHNOLOGIES AND ECONOMIC IMPORTANCE

MATERIALS SCIENCE AND TECHNOLOGIES

Additional books in this series can be found on Nova's website under the Series tab.

Additional e-books in this series can be found on Nova's website under the e-book tab.

MATERIALS SCIENCE AND TECHNOLOGIES

ADHESIVES

MECHANICAL PROPERTIES, TECHNOLOGIES AND ECONOMIC IMPORTANCE

DARIO CROCCOLO
EDITOR

New York

For permission to use material from this book please contact us:
Telephone 631-231-7269; Fax 631-231-8175
Web Site: http://www.novapublishers.com

Library of Congress Cataloging-in-Publication Data

ISBN: 978-1-63117-653-1

Published by Nova Science Publishers, Inc. † New York

CONTENTS

PREFACE

The present book is dedicated to the description of the mechanical properties, the technologies and the economic importance of adhesives.

First of all interference fitted and adhesively bonded joints has been studied as an effective means to increase the transferable load while reducing both the weight and the stress level of the joined components. In such a type of joints the influence of the assembly technique on the shear strength of the joints has been investigated. Therefore the shear strength of hybrid joints realized by 'press fitting', by 'shrink fitting' or by 'cryogenic fitting' were compared in order to suggest the best way of joining both for technological and economic importance.

Then the hybrid interference and some design aspects for practical application in (the automotive steel wheel) have been analyzed. An experimental evaluation of the contributions of the adhesive and the interference to the resultant resistance of the hybrid joint was carried out with particular attention to the phenomena occurring at the interface level and the effect of the adhesive nature, its curing technology and its mechanical response. The outcomes of the laboratory analyses were validated in the steel wheel system. The adhesive contribution mainly affects the static resistance of the hybrid joint, and is strongly related to the type of adhesive exploited. On the other hand, the interference seems to play an important role in the fatigue behavior, especially in the wheel system.

he static and fatigue strength properties of press fitted and adhesively bonded joints has been, also, studied extensively by considering the Engagement Ratio (i.e., the coupling length over the coupling diameter). Coupling and decoupling tests have been performed both on press-fitted and adhesively bonded specimens and on pin-collar samples, considering four different levels for the Engagement Ratio. The study shows that the Engagement Ratio has a negligible effect on the shear strength of the adhesive and also on the relationship between the decoupling and the coupling forces. Moreover, the obtained results show that a too high interference level in press-fitted and adhesively bonded joints may have a detrimental effect on the adhesive strength.

Furthermore the assessment of the adhesive performance of two binders for reassembling fragment porous stones and, more specifically, the effect of nano-titania in the hydration and carbonation of the derived mortars, have been investigated. The nano-titania of anatase form, has been added in mortars containing (a): binders of either lime and metakaolin or natural hydraulic lime and, (b): fine aggregates of carbonate nature. The nano-titania proportion was 4.5-6% w/w of binders. The physicochemical and mechanical properties of the nano-titania mortars were studied and compared to the respective ones of the mortars without the nano-titania addition, used as reference. DTA-TG, FTIR, SEM and XRD analyses indicated the evolution of carbonation, hydration and hydraulic compound formation during a period of one-year curing. The mechanical characterization indicated that the mortars with the nano-titania addition, showed improved mechanical properties over time, when compared to the specimens without nano-titania. The results evidenced carbonation and hydration enhancement of the mortar mixtures with nano-titania. The hydrophylicity of nano-titania enhances humidity retention in mortars, thus facilitating the carbonation and hydration processes. This property can be exploited in the fabrication of mortars for reassembling fragments of porous limestones from monuments, where the presence of humidity controls the mortar setting and adhesion efficiency. The rapid discoloration of methylene blue stains applied to mortars with nano-titania supported the self-cleaning properties of mortars with nano-titania presence. Based on the physico-chemical and mechanical characterization of the studied adhesive mortars with nano-titania, binders of metakaolin-lime and natural hydraulic lime, have been selected as most appropriate formulations for the adhesion of fragment porous stones in restoration applications.

Then the adhesive crack propagation has been investigated for some load bearing applications. Fatigue cracks, either induced by defects or by applied stresses, may appear and propagate, thus becoming potentially harmful for the structural integrity of a part or a whole structure. Therefore, in-situ structural health monitoring (SHM) of bonded joints is essential to maintain reliable and safe operaional life of these structures. This chapter presents various in-situ SHM techniques which are used to monitor bonded composite joints with a focus on fatigue crack monitoring based on the backface strain (BFS) technique. Case studies are presented and discussed for adhesively bonded single lap joint (SLJ). Sensors associated with this technique are also explained, with emphasis on Fibre Bragg Grating (FBG) optical sensors.

Finally a numerical method able to reproduce three-dimensionally the fatigue debonding and/or delamination evolution in bonded structures has been proposed in order to improve their performances. The cohesive zone model previously developed by the authors to simulate fatigue crack growth at interfaces in 2D geometries is extended to 3D cracks under mixed-mode I/II loading

The Editor
Dario Croccolo
Department of Industrial Engineering,
University of Bologna, Italy,
e-mail: dario.croccolo@unibo.it

In: Adhesives
Editor: Dario Croccolo

ISBN: 978-1-63117-653-1

Chapter 1

INFLUENCE OF THE ASSEMBLY PROCESS ON THE SHEAR STRENGTH OF SHAFT–HUB HYBRID JOINTS

D. Croccolo*[*]*, M. De Agostinis, G. Olmi and P. Mauri
University of Bologna - DIN, Bologna, Italy

ABSTRACT

Interference fitted and adhesively bonded joints, also known as hybrid joints, are an effective means to increase the transferable load while reducing both the weight and the stress level of the joined components. Many researches evaluated the strength of hybrid joints in dependence of several variables, such as the assembly pressure level, the type of materials in contact, the curing methodology, the operating temperature, and the loading type. To the authors' best knowledge, no data are available about the influence of the assembly technique on the shear strength of hybrid joints. This paper aims at filling the gap, by comparing the shear strength of hybrid joints realized by 'press fitting', by 'shrink fitting' and by 'cryogenic fitting'.

Keywords: Anaerobic adhesives, hybrid joints, shrink fit, press fit, cryogenic fit

---

[*] Correspondence information: dario.croccolo@unibo.it.

INTRODUCTION

Interference fitted and adhesively bonded joints, also known as hybrid joints (HJs), are an effective means to emphasize the transferable load while reducing both the weight and the stress level of the joined components [1]. Hybrid joints usually involve two axisymmetric members and a high strength, single component, anaerobic adhesive like LOCTITE648. Many researches evaluated the strength of HJs in dependence of several variables, such as the assembly pressure level [2,3], the type of materials in contact [4–6], the curing methodology [7], the operating temperature [8,9] and the loading type [10–12]. Interference level can be produced by different joining techniques: by driving the shaft into the hub with a standing press at R.T., by heating the hub then driving the shaft (kept at R.T.) into the hub, and finally by cooling the shaft then driving it into the hub (kept at R.T.). The first technique is called 'press fit', the other two are known as 'shrink fit' and 'cryogenic fit', respectively. To the authors' best knowledge, no experimentation has still been done to assess the static shear strength of HJs in dependence of the method by which interference is obtained. This research aims at filling the gap, by making comparisons between the shear strength of hybrid joints realized by the above mentioned assembly techniques.

METHODOLOGY

The experimentation was divided into two steps: the first part run on pin-collar specimens realized according to ISO 10123 [13], and the second one carried out on bigger sized shaft–hub couplings. The bigger sized specimens allow larger gaps for given temperature, so asto make possible the assembly process with practical temperature differentials between the parts; furthermore the bigger size allows to increase the time available for assembly operations. All the specimens were manufactured with C40 UNI EN 10083-2 steel [14], without any surface treatment. All the specimens were pushed out with the same press (Italsigma, 100 kN) and the complete force-displacement diagram was recorded. The push-out speed was set at 0.5 mm/s throughout the experimentation. Press fit specimens were assembled by means of the aforementioned press, while the clearance, shrink fit and cryogenic-fit ones were assembled manually. All the specimens assembled with adhesive cured more than 24 h at a R.T. of 20°C, in order to achieve a complete polymerization [15]. In the case of adhesively bonded joints assembled with clearance, the shear strength of the adhesive tad_cl can be calculated according to:

$$\tau_{ad\_cl} = \frac{F_{tot}}{A} = \frac{F_{ad\_cl}}{A} \tag{1}$$

where A is the coupling surface and Ftot is the push-out force. Eq. (1) holds true for clearance specimens assembled at any temperature. In the case of joints assembledwith interference and no adhesive, Eq. (2) is used, in which m is the mean coefficient of friction between the mating parts, evaluated at decoupling.

$$F_{\text{int}} = P \cdot \mu \cdot A \tag{2}$$

Therefore, the adhesive shear stress of HJs can be expressed as follows:

$$\tau_{ad\_\text{int}} = \frac{F_{tot} - F_{\text{int}}}{A} \tag{3}$$

Pin-Collar Specimens

In order to have a first glance at the performances of LOCTITE648s adhesive in dependence of the assembly temperature of the components, a first set of 13 pin-collar specimens was examined. The specimens are characterized by a mean radial clearance of 0.027 mm (standard deviation of 0.003 mm). The specimen size was chosen in accordance with ISO 10123, because it allows direct comparison with the performances stated into the datasheet of the adhesive. It was demonstrated by Croccolo et al. [12] that if only steel components are involved, the specimens can be pushed out and re-used several times without influencing the strength of the joint, provided that they are cleaned accurately after pushing out. Here, the LOCTITE7063s multi purpose cleaner was used. In the light of that, each of the 13 clearance pairs was used three times; Table 1 summarizes the test plan. It must be remarked that, in this first part of the experimentation, easily achievable heating (+180°C) and cooling (-20°C) temperatures were chosen. The choice was based on the tools readily available in the laboratory (a freezer and a small electric oven).

Table 1. Test plan for ISO 10123 pin-collar specimens, tested with clearance fits

First run	R.T. Adhesive = YES 13 tests
Second run	Collar heated at +180°C Adhesive = YES 13 tests
Third run	Pin cooled at -20°C Adhesive = YES 13 tests

Shaft-Hub Specimens

Pin-collars manufactured according to ISO 10123 have a coupling diameter of about 12.65 mm. When a shrink-fit or a cryogenic-fit has to be performed, a certain radial clearance, and a sufficient assembly time shall be ensured, before the components return to their original dimensions. It is known that a linear relationship exists between the thermal deformation of a metallic body and temperature [16].

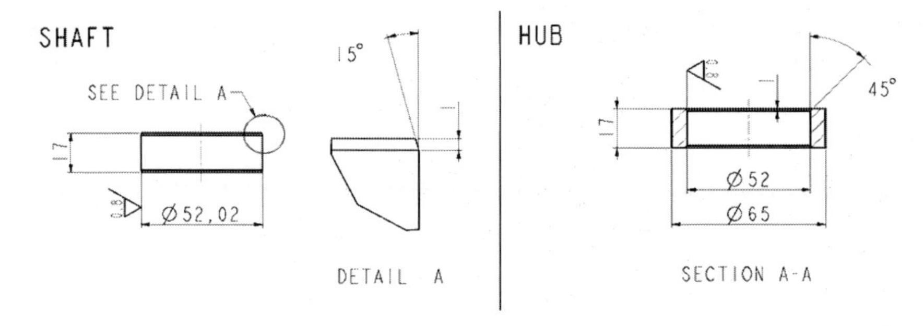

Figure 1. Shaft-hub specimen dimensions.

In the light of that, the relatively small dimensions of pin-collars realized as for ISO 10123 have a double drawback: they allow for small radial displacements, even under great temperature differentials, and they do have a small thermal inertia. In order to avoid assembly issues, the authors designed and realized the specimen shown in Figure 1, to be used in the presence of interference. The bigger size of such shaft–hub specimen, with respect to pin collars, allows for a greater assembly clearance and increases the time available for assembly operations. For example, referring to a nominal coupling diameter D_C = 52 mm, to a radial interference I=0.01 mm, to the coefficient of thermal expansion α_W=1.1 10^{-5} for steel, and to a R.T. of +20°C, a radial assembly clearance of g=0.07 mm is obtained when the hub is heated at +180°C for the shrink fit. Such assembly clearance permits to join the components by hand easily. As for cryogenic fits, by taking into account a coefficient of thermal contraction of α_C=9.5 10^{-6} and a cooling temperature for the shaft of -150°C, a radial assembly clearance of g=0.06 mm is obtained, which is still acceptable. Cryogenic fits are performed by means of liquid nitrogen and the temperatures indicated above are meant at the time of coupling. The coupling length of shaft–hub specimens was defined in order to obtain a push-out force always lower than the limit of the standing press (100 kN) whatever was the coupling condition; the material type was kept unchanged with respect to the pin-collars one. Shaft– hub specimens with interference and without adhesive were used to sample the frictional contribution Fint to be used for calculating the overall release force F_{tot}, as for Eq. (3). A total of 34 shaft–hub specimens were realized, out of which 7 were press fitted without adhesive, 7 were press fitted with adhesive, 10 were shrink fitted with adhesive and, 10 were cryogenically fitted with adhesive. In the case of shrink fit, the adhesive was dispensed on the shaft, while in the other two cases it was dispensed on the hub. Table 2 summarizes the test plan for shaft–hub couplings. Since all shaft–hub couplings were assembled with interference, the adhesive shear strength has to be calculated according to Eq. (3).

Table 2. Test plan for shaft-hub specimens, tested with interference, single run

Press-fit Adhesive = NO 7 tests	Press-fit Adhesive = YES 7 tests	Shrink-fit (Hub heated at +180°C) Adhesive = YES 10 tests	Cryogenic-fit (Shaft cooled at -150°C Adhesive = YES 10 tests

RESULTS AND DISCUSSION

Pin-Collar Specimens

The results are collected in the following tables: Table 3 refers to slip fit couplings assembled at R.T.; Table 4 refers to slip fit couplings with collars heated at +180°C; and Table 5 refers to slip fit couplings with pins cooled at -20°C. Referring to Table 3, the mean adhesive shear strength of slip fit couplings assembled at R.T. can be calculated. Assuming A=446.33 mm^2, Eq. (1) gives τ_{ad_cle}= 38:5 MPa (standard deviation 3.4 MPa), which is greater than the minimum indicated by the manufacturer in the product data-sheet (τ_{ad_min}= 25 MPa). A picture of the pushed out specimens is given in Figure 2: the polymerised adhesive residue is clearly visible on both the mating surfaces, indicating the expected cohesive failure mode.

Figure 2. Slip fit coupling assembled at R.T.

Slip fits performed by heating the collars at +180°C produced the results reported in Table 4; the mean adhesive shear strength is here equal to 19.5 MPa (standard deviation 4.3 MPa), which is lower than the minimum suggested by the data sheet of the adhesive. This occurrence can be explained by a certain amount of rust observed on the collar coupling surface, due to heating. Rust is, for anaerobic products, an inert substrate which inhibits the complete polymerization of the adhesive and, therefore, reduces the adhesive performances. Figure 3 shows a cohesive failure mode, while some rust spots are visible on the annular surface of the collar

Figure 3. Slip fit coupling assembled after heating the collar at +180°C.

Figure 4. Slip fit coupling assembled after cooling the pin at -20°C.

Slip fits performed by cooling the pins at -20°C produced the results reported in Table 5; the mean adhesive shear strength drops down to 5.7 MPa (standard deviation 1.8 MPa), which is less than one fourth of the minimum suggested by the data sheet. That outcome can be explained by a large presence of moisture on the pin surface during assembly. In fact, the frost present on the pin as extracted from the freezer, transforms into moisture in a short time at R.T. (20°C). This contamination, well highlighted in Figure 4, drastically inhibits polymerization, reducing the adhesive shear strength.

Table 3. Results for pin-collar specimens assembled with clearance at R.T.

ID	Radial clearance [mm]	Coupling area [mm^2]	Push out force [N]	τ_{ad_cl} [MPa]
1c	0.024	423.5	14,000	33.1
2c	0.025	442.59	19,500	44.1
3c	0.023	442.69	17,500	39.5
4c	0.027	442.83	18,200	41.1
5c	0.020	442.09	16,100	36.4
6c	0.029	445.15	17,200	38.6
7c	0.031	442.66	17,600	39.8
8c	0.024	442.73	18,500	41.8
9c	0.031	442.11	18,100	40.9
10c	0.030	443.46	15,900	35.9
11c	0.025	443.19	14,200	32
12c	0.031	442.97	17,300	39.1
13c	0.020	443.29	16,800	37.9

Table 4. Results for pin-collar specimens assembled with clearance and heating the hub

ID	Radial clearance [mm]	Coupling area [mm2]	Push out force [N]	τ_{ad_cl} [MPa]
1c	0.024	423.5	9,800	23.1
2c	0.025	443.46	9,700	21.9
3c	0.023	442.83	12,500	28.2
4c	0.027	442.69	6,800	15.4
5c	0.020	443.29	8,700	19.6
6c	0.029	445.15	7,000	15.7
7c	0.031	442.66	8,500	19.2
8c	0.024	442.73	8,200	18.5
9c	0.031	442.93	5,800	13.1
10c	0.030	442.11	9,600	21.7
11c	0.025	443.19	10,100	22.8
12c	0.031	442.97	9,000	20.3
13c	0.020	442.09	5,900	13.3

Table 5. Results for pin-collar specimens assembled with clearance and cooling the shaft

ID	Radial clearance [mm]	Coupling area [mm2]	Push out force [N]	τ_{ad_cl} [MPa]
1c	0.024	423.5	2,500	5.9
2c	0.025	443.46	3,000	6.8
3c	0.023	442.83	3,100	7.0
4c	0.027	442.69	2,300	5.2
5c	0.020	443.29	2,000	4.5
6c	0.029	445.15	3,000	6.7
7c	0.031	442.66	1,700	3.8
8c	0.024	442.73	1,900	4.3
9c	0.031	442.93	1,500	3.4
10c	0.030	442.11	3,800	8.6
11c	0.025	443.19	3,400	7.7
12c	0.031	442.97	3,300	7.4
13c	0.020	442.09	1,400	3.2

Shaft-Hub Specimens

The results are collected in the following tables: Table 6 refers to press-fit couplings assembled at R.T. without adhesive, while Table 7 refers to press fit couplings assembled at R.T. and reinforced with adhesive. Table 8 refers to shrink fit couplings with adhesive and hubs heated at +180°C and Table 9 refers to cryogenic fit couplings with adhesive and shafts cooled at -150°C. From suitable elaboration of the data relevant to press-fit couplings assembled at R.T., it is possible to evaluate the mean coefficient of friction between the mating surfaces during disassembly, which, by assuming an engaged area $A=2,454\ mm^2$, is $\mu=0.207$ (standard deviation 0.035). Then, the frictional contribution for a given coupling can be calculated with Eq. (2). Referring, for example, to press-fit shaft hub specimens (Table 7), it can be calculated a mean adhesive shear strength of $\tau_{ad\_mean}=22.9$ MPa with a standard deviation of 3.4 MPa (mean coupling pressure of 5.4 MPa). Both the friction coefficient and the τ_{ad_mean} values are aligned with the data shown in [12] for steel–steel couplings. It is confirmed that, in the case of steel–steel couplings, the frictional contribution to the overall push-out force of HJs is negligible if compared to the contribution of the adhesive. As for the shrink-fit process, referring to Table 8, a mean coupling pressure p=15.8 MPa can be calculated according to Croccolo et al. [1]. Therefore, by assuming a friction coefficient μ=0.207 and an engaged area A=2,454 mm^2, the frictional contribution can be calculated as for Eq. (2). Then, a mean adhesive shear strength of τ_{ad_mean}=34.7 MPa can be retrieved, with a standard deviation 2.8 MPa. Such a shear strength is greater than the minimum suggested by the data sheet of the adhesive as well as greater than that shown above for pin-collar specimens, heated and with no interference. In fact, it has been shown recently by Castagnetti and Dragoni [17] that anaerobic adhesives increase their strength if combined with a moderate coupling pressure (a moderate interference). A possible explanation to this occurrence is that a thin film of adhesive is retained in the surface roughness of the mating parts, regardless of the coupling pressure level: in analogy to what happens for regular polymers, if high local pressure is applied, that film can build up a greater shear strength than that at ambient pressure [18]. Figure 5 shows the fracture surface of two shrink-fit shaft–hub specimens: the failure mode is again cohesive, and the adhesive residue is uniformly distributed over the whole mating surfaces.

Figure 5. Shaft–hub specimen assembled by shrink fitting at +180°C, after pushout.

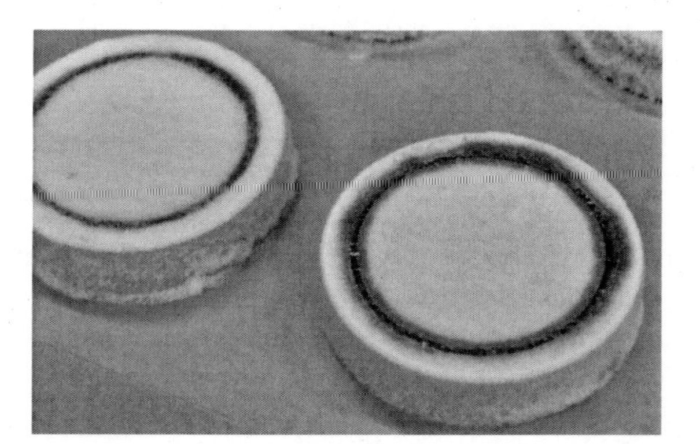

Figure 6. Shaft–hub specimens assembled by cryogenic fitting at -150°C, just after coupling.

As for cryogenic fitted shaft–hub specimens a mean coupling pressure of 8.6 MPa can be calculated from data retrieved by Table 9. The mean adhesive shear strength is therefore τ_{ad_mean}=13.2 MPa, with a standard deviation of 4.6 MPa. Such shear strength is lower than the minimum suggested by the data sheet of the adhesive. On the contrary, if compared to pin-collar specimens assembled with clearance after cooling the collar at -20°C, the result in terms of adhesive shear strength is higher in the presence of interference, which confirms what was stated above about the beneficial effect of coupling pressure. After cryogenic fit assembly, a considerable amount of frost builds up on the specimen surface (Figure 6). '

Figure 7. Fracture surface of a shaft–hub coupling assembled by cryogenic fitting at -150°C.

Figure 7 shows the fracture surface of a cryogenic fitted shaft–hub specimen. The surface is quite inhomogeneous, with some areas characterized by adhesive failure and others showing cohesive failure. It can be appreciated from Figure 7 that the frost flourished on the specimens causes a growth of rust on the surfaces.

The remarks mentioned above are summarized in Figs. 8–10. Figure 8 shows the full push-out stroke for three shaft–hub specimens: ID5 (dash-dot line) assembled by shrink fit, ID29 (dashed line) assembled by cryogenic fit, and ID23 (solid line) assembled by press fit with adhesive. The assembly interference was similar for the three specimens (see Tables 7–9), therefore its contribution to the overall strength of the joints can be considered equal among the three tests. Moreover, it can be appreciated how the initial elastic part of the push-out curves is quite different among the three tests, because it is mostly influenced by the adhesive strength. On the contrary, the subsequent descending part related to the frictional contribution, is similar for the three curves.

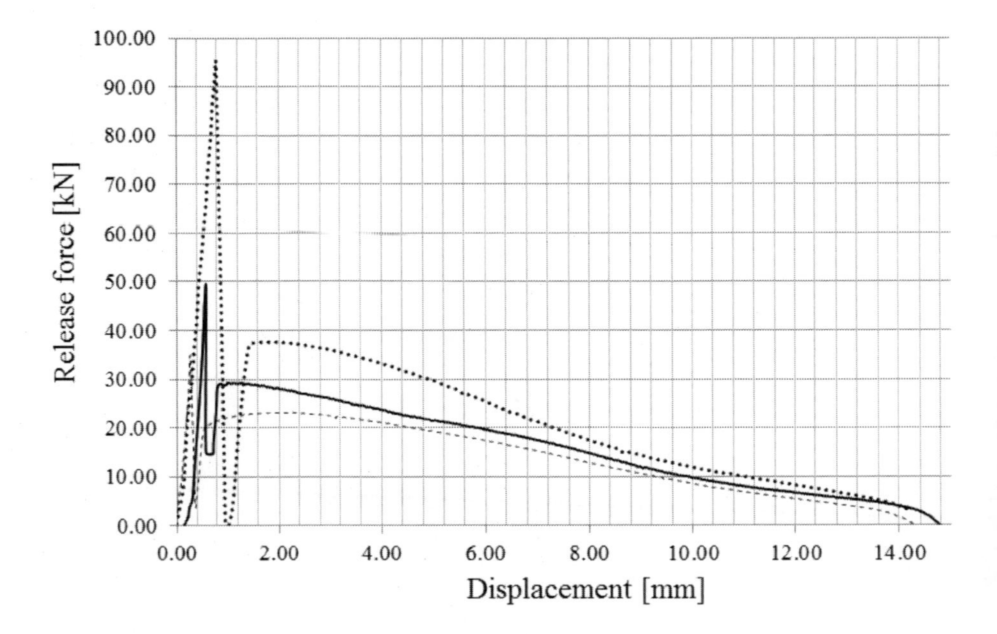

Figure 8. Comparison of static shear strength of shrink fit ID5 (dotted line), cryogenic fit ID29 (dashed line) and press fit ID23 (solid line) specimens.

Figures 9 and 10 report histograms relevant to the shear strength of the different joints, taking into account the sole adhesive contribution and the overall shear strength of the joint, respectively. Looking at Figure 9, it can be appreciated how the joints of bigger size (shaft–hub specimens) assembled by press-fitting at R.T., have a lower shear strength than pin-collar specimens assembled with clearance at R.T. This occurrence may be explained by observing that, during the press fit operation, a certain amount of adhesive is stripped away from the mating surfaces, thus reducing the actual bonded area. Moreover, Figure 9 shows that both shrink fit and cryogenic fit shaft–hub joints have a higher adhesive shear strength than pin-collars assembled with comparable temperature differentials. It must be remarked that shaft–hub joints are assembled with interference while pin-collars are assembled with clearance. To some extent, this occurrence confirms the beneficial effect of pressure on the adhesive shear strength. Finally, looking at the left side of Figure 9, it can be seen that both shrink fitting and cryogenic fitting determine a decrease of the shear strength of the adhesive with respect to clearance fitting at R.T. The effect is much more pronounced in the case of cryogenic fitting. Figure 10 shows how, in the case of steel on steel HJs, the frictional contribution to the overall shear strength of the joint is negligible with respect to the adhesive contribution.

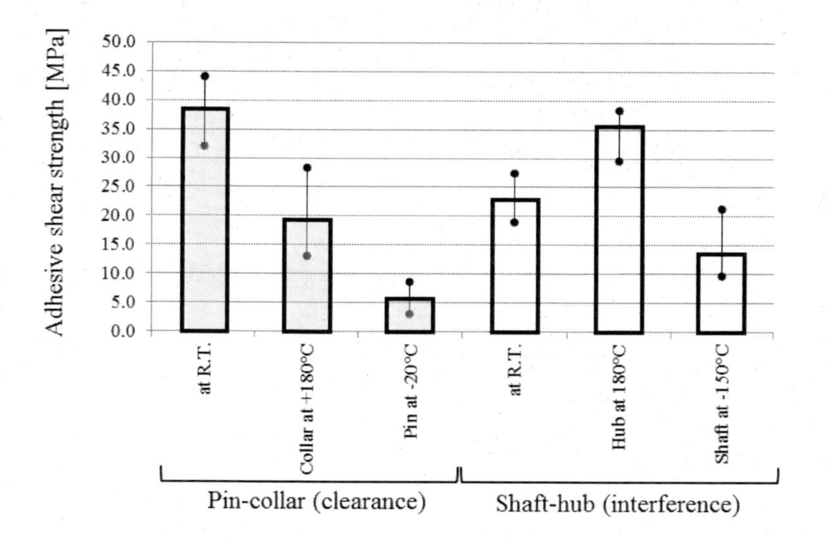

Figure 9. Adhesive shear strength: clearance fits vs. interference fits.

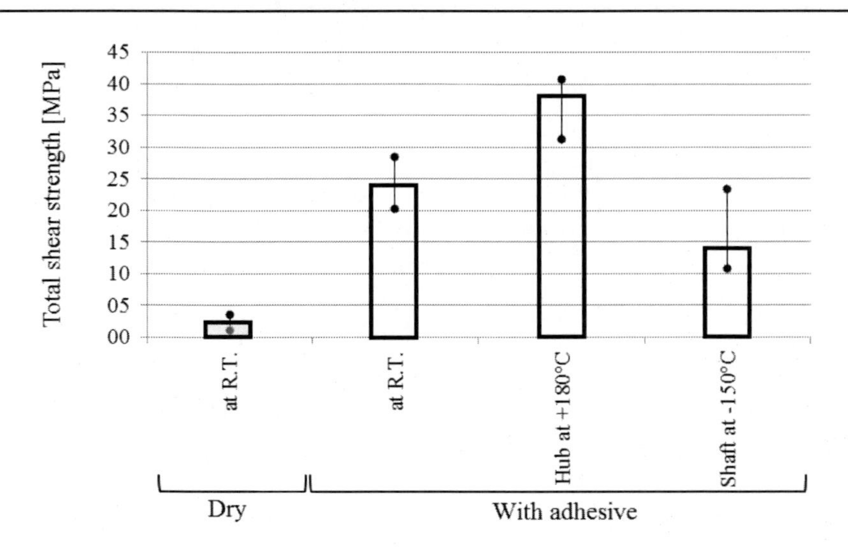

Figure 10. Total shear strength: dry vs. adhesively bonded shaft–hub specimens.

Table 6. Results for shaft–hub specimens assembled with interference and without adhesive

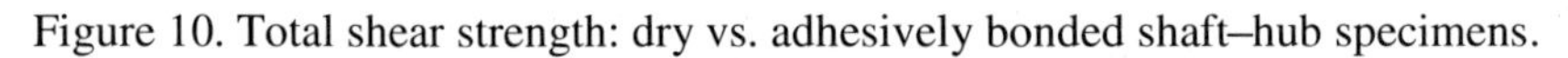

ID	Radial interference [mm]	Coupling area [mm^2]	Push out force [N]
32	0.010	2,448.43	5,631
33	0.007	2,464.93	2,758
34	0.012	2,455.20	4,972
35	0.013	2,461.66	8,798
36	0.010	2,451.70	4,933
37	0.010	2,443.68	4,390
38	0.010	2,460.19	3,769

Table 7. Results for shaft–hub specimens assembled by press fitting with adhesive

ID	Radial interference [mm]	Coupling area [mm^2]	Push out force [N]	τ_{ad_int} [MPa]
17	0.005	2,453.88	52,000	20.70
23	0.007	2,453.33	50,000	19.40
28	0.007	2,453.33	65,000	25.52
13	0.007	2,455.44	60,000	23.46
21	0.007	2,455.13	65,000	25.50
31	0.008	2,455.05	70,000	27.05
19	0.01	2,456.44	50,000	18.41

Table 8. Results for shaft–hub specimens assembled by shrink fitting with adhesive

ID	Radial interference [mm]	Coupling area [mm^2]	Push out force [N]	τ_{ad_int} [MPa]
1	0.010	2,451.86	91,230	34.29
2	0.012	2,454.89	100,000	37.33
3	0.010	2,461.66	99,010	37.30
4	0.010	2,455.13	89,070	33.36
5	0.008	2,453.73	94,950	36.26
6	0.008	2,455.20	76,780	28.84
7	0.013	2,453.49	100,000	36.85
24	0.017	2,435.04	90,750	32.40
9	0.012	2,454.73	90,240	33.35
10	0.012	2,455.36	100,000	37.32

Table 9. Results for shaft–hub specimens assembled by cryogenic fitting with adhesive

ID	Radial interference [mm]	Coupling area [mm^2]	Push out force [N]	τ_{ad_int} [MPa]
18	0.003	2,455.60	33,690	12.74
8	0.005	2,452.41	31,130	11.24
30	0.005	2,451.93	46,010	17.30
20	0.005	2,455.36	47,430	17.86
22	0.005	2,455.36	28,430	10.12
29	0.005	2,453.73	34,360	12.54
16	0.005	2,452.25	26,460	9.33
26	0.007	2,453.33	30,660	10.55
27	0.010	2,454.65	32,060	10.14
11	0.010	2,456.44	57,540	20.50

CONCLUSION

The assembly technique does influence the final strength of the joint, since especially cryogenic fitting determines an appreciable decay of the adhesive shear strength, even if the components are assembled with interference. The negative effect of heating at +180°C is less severe. Moreover, the shear strength of cylindrical joints reinforced with strong anaerobic adhesive increases with coupling pressure. In fact, comparing the performances of slip fit joints with those of interference fit joints, both reinforced with adhesive and assembled by shrink fit and cryogenic fit, interference fitted joints systematically overperform clearance fit joints in terms of adhesive shear strength. In the light of the results shown above, in practical applications, where heating is always associated with interference, it is expected that the beneficial effect of contact pressure could counterbalance the negative effect of heating. Press fit joints deserve particular attention, because they show a slightly lower shear strength than slip fit ones assembled at R.T. Lower results are observed on press fit joints with adhesive since the adhesive is stripped away from the mating surfaces during the assembling while the beneficial effect of coupling pressure increases the performances of the remaining adhesive; the overall performance of the joint is, therefore, comparable to that of slip fit joints reinforced with adhesive. As an extension of the present work, the effect of the ratio between the coupling length and the coupling diameter on the adhesive performances will be investigated in the future.

REFERENCES

[1] Croccolo, D., De Agostinis, M., Vincenzi, N., 2012. Design and optimization of shaft-hub hybrid joints for lightweight structures: analytical definition of normalizing parameters. *Int. J. Mech. Sci.* 56 (1), 77-85.

[2] Dragoni, E., Mauri, P., 2000. Intrinsic static strength of friction interfaces augmented with anaerobic adhesives. *Int. J. Adhes. Adhes.* 20 (4), 315-321.

[3] Sekercioglu, T., Gulsoz, A., Rende, H., 2005. The effects of bonding clearance and interference fit on the strength of adhesively bonded cylindrical components. *Mater. Des.* 26 (4), 377-381.

[4] Sekercioglu, T., 2005. Shear strength estimation of adhesively bonded cylindrical components under static loading using the genetic algorithm approach. *Int. J. Adhes. Adhes.* 25 (4), 352-357.

[5] Sekercioglu, T., Meran, C., 2004. The effects of adherend on the strength of adhesively bonded cylindrical components. *Mater. Des.* 25 (2), 171-175.

[6] Croccolo, D., De Agostinis, M., Vincenzi, N., 2012. Design of hybrid steel-composite interference fitted and adhesively bonded connections. *Int. J. Adhes. Adhes*. 37, 19-25.

[7] Mengel, R., Haeberle, J., Schlimmer, M., 2007. Mechanical properties of hub/shaft joints adhesively bonded and cured under hydrostatic pressure. *Int. J. Adhes. Adhes*. 27 (7), 568-573.

[8] Da Silva, L.F.M., Adams, R.D., 2007. Joint strength predictions for adhesive joints to be used over a wide temperature range. *Int. J. Adhes. Adhes*. 27 (5), 362-379.

[9] Da Silva, L.F.M., Adams, R.D., 2007. Adhesive joints at high and low temperatures using similar and dissimilar adherends and dual adhesives. *Int. J. Adhes. Adhes*. 27 (5), 216-226.

[10] Adams, R.D., Comyn, J., Wake, W.C., 1977. Structural adhesive joints in engineering. London: Chapman & Hall.

[11] Croccolo, D., De Agostinis, M., Vincenzi, N., 2011. Experimental analysis of static and fatigue strength properties in press-fitted and adhesively bonded steel-aluminium components. *J. Adhes. Sci. Technol*. 25 (18), 2521-2538.

[12] Croccolo, D., De Agostinis, M., Vincenzi, N., 2010. Static and dynamic strength evaluation of interference fit and adhesively bonded cylindrical joints. *Int. J. Adhes. Adhes*. 30 (5), 359-366.

[13] ISO 10123, 1990. Adhesives—determination of shear strength of anaerobic adhesives using pin-and-collar specimens.

[14] UNI EN 10083-2, 2006. Steels for quenching and tempering—part 2: technical delivery conditions for non alloy steels.

[15] Loctite, 2005. 648 Technical Data Sheet.

[16] Niemann, G., Winter, H., Hohn, B.R., 2005. Maschinenelemente. Berlin: Springer- Verlag.

[17] Castagnetti, D., Dragoni, E., 2012. Predicting the macroscopic shear strength of adhesively bonded friction interfaces by microscale finite element simulations. *Comput. Mater. Sci*. 64, 146-150.

[18] Raghava, R., Caddell, R., Yeh, G., 1973. The macroscopic yield behaviour of polymers. *J. Mater. Sci*. 8 (2), 225-232.

In: Adhesives
Editor: Dario Croccolo

ISBN: 978-1-63117-653-1

Chapter 2

THE HYBRID PRESS-FITTED ADHESIVE BONDED JOINT: THEORETICAL ASPECTS AND A PRACTICAL APPLICATION ON THE AUTOMOTIVE WHEEL

Giorgio Gallio*[*], *Mariangela Lombardi*, *Paolo Fino and Laura Montanaro
Politecnico di Torino, Department of Applied Science and Technology, Torino, Italy

ABSTRACT

Nowadays, adhesive bonding is increasingly exploited in industrial applications, especially in multi-material lightweight structures, for its ability to join dissimilar materials together. In order to overcome the limits related to the durability of adhesives exposed to harsh environmental conditions, adhesive bonding can be used in combination with other joining techniques. This chapter examines the hybrid interference fitted-adhesive bonded joint and some design aspects for its practical application in a case study: the automotive steel wheel. An experimental evaluation of the contributions of the adhesive and the interference-fit to the resultant resistance of the hybrid joint was carried out with particular attention to the phenomena occuring at the interface

[*] Corresponding Author address: Email: gallio.giorgio@gmail.com.

level and the effect of the adhesive nature, its curing technology and its mechanical response. The outcomes of the laboratory analyses were validated in the steel wheel system. It was found that the adhesive contribution mainly affected the static resistance of the hybrid joint, and it was strongly related to the type of adhesive exploited. On the other hand, the interference-fit seemed to play an important role in governing the fatigue behaviour, especially in the wheel system.

Keywords: Interference fit, hybrid interface, mechanical properties of adhesives, joint design

INTRODUCTION

Adhesive bonding is a suitable technology to join dissimilar materials together and recently its exploitation has greatly increased in industry for bonding new hybrid lightweight structures (Amancio-Filho & Santos, 2009; Dillard, 2010; Petrie, 2007). The main drawback of adhesives is related to their durability in harsh conditions due to their polymeric nature. In fact, especially in outdoor environments, adhesive joints are susceptible to water and humidity (Adams, Comyn, & Wake, 1997; Kinloch, 1990; Petrie, 2007). Nevertheless, adhesive bonding can be used in combination with other traditional joining methods generating a hybrid joint with the aim of combining the advantages of the different techniques and trying to overcome their drawbacks (da Silva, Pirondi, & Ochsner, 2011).

The interference fit is a common technique to join cylindrical parts together. The most common joint geometry consists in a shaft fitted into a hub or a ring. In order to guarantee the coupling, an *interference* δ exists among the hub and the shaft.

The pull-out strength of a solely interference fit joint depends on the *pressure* P between the hub and the shaft, the *friction coefficient* μ, and the *contact area* A, according to the equation (Croccolo, De Agostinis, & Vincenzi, 2010; Croccolo, de Agostinis, & Vincenzi, 2011; Kleiner & Fleischmann, 2011):

$$F_{int} = \mu P A \tag{1}$$

Moreover, the radial pressure P is a linear function of the interference level according to the Lamè's thick walled cylindrical theory (Croccolo et al., 2010; Kleiner & Fleischmann, 2011).

In the hybrid interference fitted-adhesive bonded joint the presence of an adhesive into an existing interference fit design can lead to a considerable strength enhancement. Moreover, the adhesives fill the roughness of the mating surfaces bringing the real contact area between the two fitted components to 100 %, thus improving the stress distribution in the joint area (Abenojar, Ballesteros, del Real, & Martinez, 2013; Castagnetti & Dragoni, 2012; Dragoni & Mauri, 2002).

Considering that this hybrid joint is a combination of the adhesive bonding and the mechanical tightening based on friction, the first main object of the technical research was focused on the comprehension of the interaction of these two techniques. A useful method developed for this aim is based on the concept of the superposition of the effects (Croccolo et al., 2011; Dragoni & Mauri, 2000; Kleiner & Fleischmann, 2011). According to this idea, the resistance of the hybrid joint is provided by the sum of the contributions due to the friction forces and the adhesive ones along the joint area, as two variables computed independently, according to the equation:

$$F_{hyb} = (\mu P + \tau_a f) \cdot A \tag{2}$$

where τ_a is the *static resistance* of the adhesive and f is a *correction factor* dependent on different parameters such as the assembly technique, the geometry of the joint and the adherend materials.

It is important to underline that the superposition approach has been not always confirmed at the experimental level. In some cases, in fact, the total resistance of the hybrid joint increased as the pressure between the components rised, but with a different rate (Dragoni & Mauri, 2000, 2002; Sawa, Yoneno, & Motegi, 2001; Yoneno, Sawa, Shimotakahara, & Motegi, 1997). Dragoni and Mauri (Dragoni & Mauri, 2000, 2002) developed a micro-mechanical model able to explain this different behavior in the case of strong and weak anaerobic adhesives subjected to a clamping pressure. In addition, recently, a numerical model has been developed and validated on the basis of this micro-mechanical model (Castagnetti & Dragoni, 2012, 2013).

The static and dynamic mechanical behavior of the hybrid interference fitted-adhesive bonded joint has been investigated from the numerical and experimental point of view by various authors, taking into consideration the effect of different important parameters such as: the pressure between the components (Croccolo et al., 2010, 2011; Dragoni, 2003; Mengel, Häberle, & Schlimmer, 2007; a. Oinonen & Marquis, 2011; A. Oinonen & Marquis, 2011; Sawa et al., 2001; Sekercioglu, Gulsoz, & Rende, 2005; Sekercioglu, 2005;

Yoneno et al., 1997), the assembling technique employed (Croccolo, De Agostinis, & Mauri, 2013), the fitted position (Kawamura, Sawa, Yoneno, & Nakamura, 2003), the different materials involved (Croccolo et al., 2010, 2011; Mengel et al., 2007), the roughness of the surface (A. Oinonen & Marquis, 2011; Sekercioglu, 2005) and the use of either epoxy and anaerobic adhesives (Mengel et al., 2007; A. Oinonen & Marquis, 2011).

Generally acrylic anaerobic adhesives, also known as "retaining compounds", are employed in interference fitted-adhesive bonded cylindrical joints, particularly for their suitable curing technology. Activated by metal ions present on the adherends, anaerobic adhesives are able to cure in absence of oxygen, as it happens in a tightly closed joint (Comyn, 1997; Moane, Raftery, Smyth, & Leonard, 1999; Petrie, 2007). In fact, anaerobic acrylics have been effectively used as blockers and sealants, to improve the performance of mechanical tightening joints especially in the automotive industry (Dragoni & Mauri, 2000; O'Reilly, 1990). Nevertheless, in certain particular industrial applications other curing technologies could be more appropriate, especially in more complex joint geometries where both interference and clearance zones are included or different mechanical properties have to be matched. Considering these further potential applications, it should be important to understand the effect of the adhesive chemical nature, curing technology and mechanical properties on the performance of the hybrid joint.

This chapter completes the experimental results described in previous works (Gallio et al., 2014; Gallio et al., 2013) in order to better understand the effect of adhesive chemical nature, curing technology and mechanical properties on the press-fitted hybrid joint, and to further investigate the phenomena acting at the interference level and involving parameters related to tribology and rheology.

Finally, a case study describing the application of the hybrid interference fitted-adhesive bonded joint technique on the automotive steel wheel system is presented. The study was carried out in collaboration with the MW company, a division of the CLN group. The business of MW concerns the steel wheel market for passenger cars, light and heavy commercial vehicles and motorcycles. Bonded wheel prototypes were submitted to static and dynamic mechanical tests adopted for components validation. The results obtained on bonded wheel prototypes were compared with the traditional welded wheels and with the samples tested at laboratory level.

METHODS

In order to test the behavior of different adhesive systems in the hybrid joint and to analyze the influence of the interference contribution, hub/shaft laboratory samples were created.

The hollow tubes and the shafts were designed according to the requirements imposed by our traction and fatigue test machines. The laboratory samples were machined in C40 steel, the nominal coupling diameter of the samples was set to 30 mm and the coupling length was fixed to 10 mm (Gallio et al., 2014, 2013). A picture of the components is reported in figure 1 (Gallio et al., 2013).

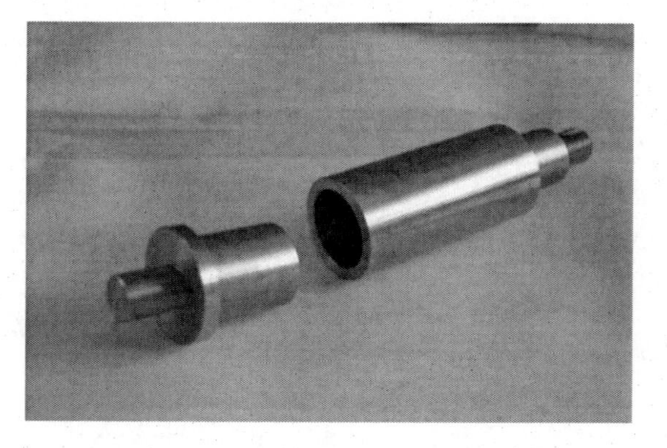

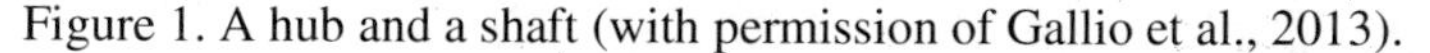

Figure 1. A hub and a shaft (with permission of Gallio et al., 2013).

Generally, the two cylindrical components could be assembled through two principal methods: the shrink-fit and the press-fit. In the case of the shrink-fit, the preferred approach is to apply the adhesive to the male component and to heat the female to obtain the needed fitting clearance. In the case of the press-fit assembling technique, the adhesive is generally applied as a film to one or both substrates and the parts are coupled by means of a standing press with possible problems concerning misalignments and the splitting away fo the adhesives (Croccolo et al., 2013; Kleiner & Fleischmann, 2011).

In our experiments, the hub/shaft joints were assembled according to the press-fit method. The alignment between the shaft and the hub was a crucial parameter to obtain repeatable joints. A not-axial assembling induces far higher coupling and decoupling loads than an axial one. A method of guides was then exploited to guarantee the alignment of the couples (more information can be found in previous papers (Gallio et al., 2014, 2013)). The pressing was performed by means of an automatic press in order to assure the same pressing rate for each sample. During the press coupling operation, the samples (both hub and shaft) were not fastened but self-arranged into the guides. The alignment was confirmed by putting in rotation the assembled joints and measuring the eccentricity between the hub and the shaft.

The joint geometry of our samples surely implied the presence of spew fillets, being the shaft press-fitted into the hub. The external spew fillet was visible after the coupling and it was removed before the adhesive cure. The internal one was observed after the breaking of the joint, but the geometry of the sample made impossible to remove it. Indeed, according to the pin and collar test method, a cord shaped adhesive spew must remain in one of the collar ends (Abenojar et al., 2013).

In order to evaluate the performances of the interference fitted-adhesive bonded joints, the samples were tested under an axial pull-out load.

The traction test was carried out at a crosshead speed of 1.3 mm/min. Considering that no standards are available for the hybrid interference fitted joint, the crosshead speed was selected on the basis of the low load rate generally suggested in the standard shear tests on adhesive joints, in particular referring to the load provided per unit area per second (Abenojar et al., 2013; Astm D1002-01, 2001; Astm D5656-01, 2001).

Two sets of experiments were carried out separately on different test machines. The first was related to test different adhesives in the interference fit system (Gallio et al., 2013). In this experimental part, the interference was set to a coupling condition of H7/p6 according to the ISO system of tolerances (ISO 286-1:2010, 2010). The tested adhesives belong to the adhesive families most employed for structural purposes, based on the Epoxy, Acrylic and Polyurethane systems (Comyn, 1997; Dillard, 2010; Petrie, 2007). The different adhesive families can be classified on the basis of the adhesion strength that they provide against their flexibility. Generally rigid adhesives provide high joint strengths and are selected for applications requiring high shear resistance and low peel and impact strength. On the other hand, flexible adhesives are characterized by a lower adhesion strength, but are suited for joints loaded in peel or cleavage, or in presence of components vibrations, and they usually provide a good impact resistance (Adams et al., 1997; Petrie, 2007).

The investigated adhesives were: **(1)** (*R-EP*) a *2k* rigid epoxy adhesive, Hysol® 9492, with high resistance (Bulk tensile strength = 31 MPa), but low flexibility (Elongation at break = 0.8 %) (Henkel Ag & Co. KGaA., 2011); **(2)**

(*FT-EP*) a *2k* flexibilized and toughened epoxy adhesive, Scotch-weld® DP490, with high resistance (Shear on grit blasted steel = 28.7 MPa) and high peel resistance (Peel strength = 9,2 N/mm) (3M Company, n.d.); **(3)** (*AC*) a *1k* high resistant anaerobic acrylic adhesive, Loctite® 620 (Shear on steel = 17.2 MPa) (Henkel Ag & Co. KGaA., 2011); **(4)** (*PU*) a *2k* modified polyurethane adhesive, Araldite® 2029, that provides characteristics of both resistance (Shear on grit blasted steel = 24 MPa) and flexibility (Elongation at break = 39 %) (Huntsman Corporation, 2013).

The design of the experiment is depicted in table 1, a minimum of 4 specimens were tested for each sample type.

Table 1. Design of the experiment for different adhesives in the interference fit

Interference H7/p6	No adhesive	Bonded with *R-EP*	Bonded with *FT-EP*	Bonded with *AC*	Bonded with *PU*
Clearance H7/f7	/	Bonded with *R-EP*	Bonded with *FT-EP*	Bonded with *AC*	Bonded with *PU*

The second set of experiments was carried out to identify the interference contribution in the hub/shaft samples and its interaction in the hybrid joints bonded with the *FT-EP* adhesive (Gallio et al., 2014). The design of the experiment is depicted in table 2, a minimum of 4 specimens were tested for each sample type.

Table 2. Design of the experiment for different interference levels in the hybrid joint

Bonded with *FT-EP*	Nominal δ -22.5 μm	Nominal δ 10 μm	Nominal δ 22.5 μm	Nominal δ 35 μm
Not Bonded	/	Nominal δ 10 μm	Nominal δ 22.5 μm	Nominal δ 35 μm

Fatigue tests on hybrid joints bonded with the *FT-EP* adhesive were also carried out, by using a tension-tension fatigue test (Croccolo et al., 2010; Sekercioglu et al., 2005) on a resonant testing machine.

Some considerations about the tension-tension variable amplitudes employed has to be done. The minimum load amplitude was set constant for each sample, on the basis of the resistance provided by *the interference contribution* (F_{int}). The maximum amplitude was varied among different percentages of the static total resistance of the hybrid joint (F_{hyb}). Indeed, Dragoni et al. (Dragoni, 2003) showed that the strength decay of hybrid joints subjected to fatigue loading could be imputable to the adhesive contribution, meanwhile the non-bonded interface is virtually unaffected by a fatigue decay. A schematic example of the fatigue cycles performed is presented in figure 2.

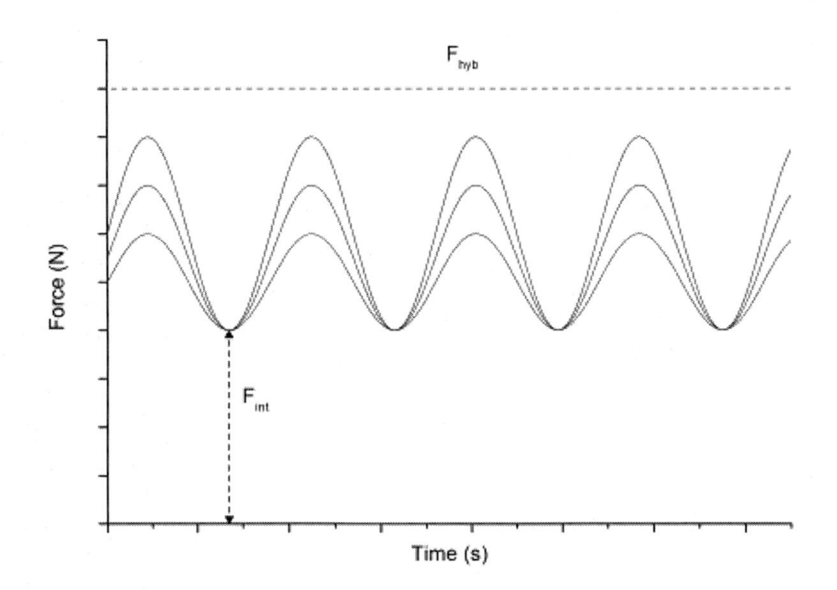

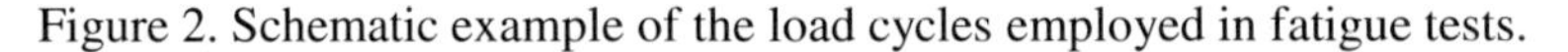
Figure 2. Schematic example of the load cycles employed in fatigue tests.

The load cycle frequency of the fatigue test was dependent on the typical frequency of the samples as the used fatigue machine operates in resonance frequency. The single fatigue test was considered terminated at the attainment of 10^6 fatigue cycles or at the joint failure (identified by a variation in the testing resonance frequency). Two specimens were tested for each investigated amplitude.

THE INTERFERENCE CONTRIBUTION

Unbonded interference-fit samples were prepared by using the press-fitted hub/shaft joints and tested in static axial pull-out in order to understand the strength contribution of the interference-fit alone (Gallio et al., 2014). Assuming the same friction coefficient between steel and steel for all the samples, a linear relationship between the interference and the pull-out force according to the theoretical equation (eq.1) should be expected at the experimental level. Nevertheless, the experimental results related to the unbonded joints were affected by a wide variability of the results. The ideal system described by the equation did not consider the phenomena of wear and micro-adhesion that took place in the samples between the two steel components, as illustrated in figure 3 for both the sliding bodies.

In order to quantify the extent of the wear on the decoupled samples, their surface was analyzed by means of a profilometer. By subtracting the measured real profile to the averaged ideal one, an absolute value of discrepancy was obtained. This discrepancy parameter was labeled as "scratch profile" and it indicates the extent of the wear phenomena (Gallio et al., 2014).

The importance of the wear and micro-adhesion phenomena on the resistance of the unbonded joints could be appreciated by plotting the measured pull-out loads as a function of the scratches profile values (Figure 4) (Gallio et al., 2014).

It can be noted that the extent of the damage caused by wear phenomena was higher in the unbonded interference samples that showed stronger resistance to the decoupling. In fact, when a metallic junction is formed between two sliding surfaces made of the same metal, the wear phenomena including processes of deformation, local adhesion and welding can appreciably increase its shear-strength (Bowden & Tabor, 1950).

Because of this phenomena, it was not possible to detect the interference contribution by the unbonded interference joints.

a) 

b) 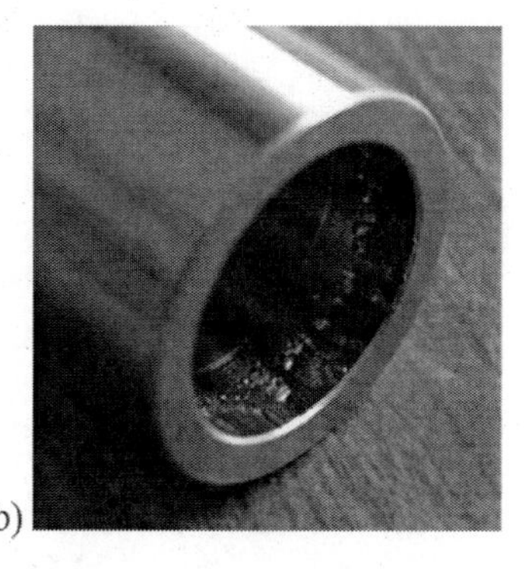

Figure 3. Sample at high interference level characterized by large surface damages after the decoupling: a) shaft and b) hub (with permission of Gallio et al., 2014).

Differently, no scratches and wear were present in the case of the hybrid joints after the decoupling, being their mating area covered by a thin layer of adhesive, as illustrated in figure 5. During the coupling operation, the adhesive, still liquid, acts as a lubricant preventing friction effects. During the decoupling, the cured adhesive avoids the direct metal-metal contact and consequently the formation of scratches (Gallio et al., 2014).

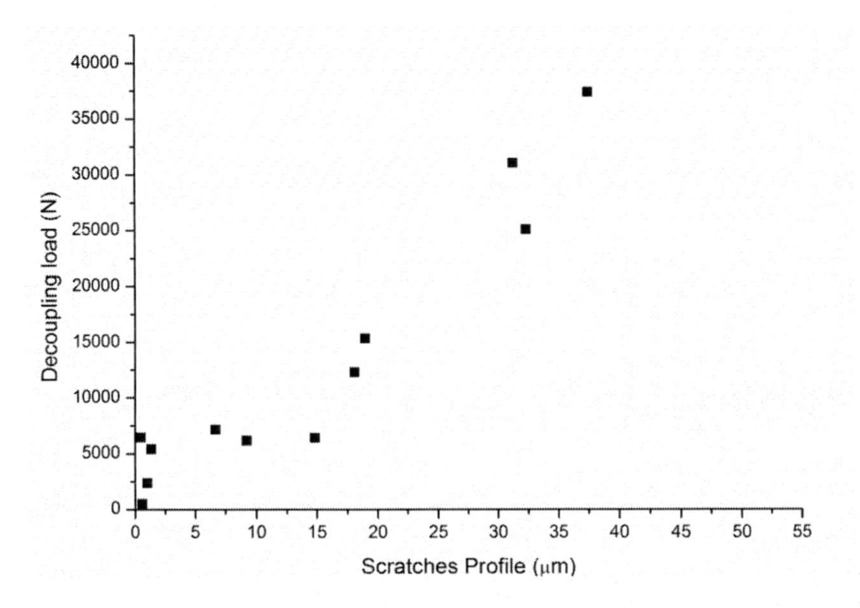

Figure 4. Relation of the scratches profile, consequence of the wear and micro-adhesion metal to metal phenomena, with the resistance of the interference joint (with permission of Gallio et al., 2014).

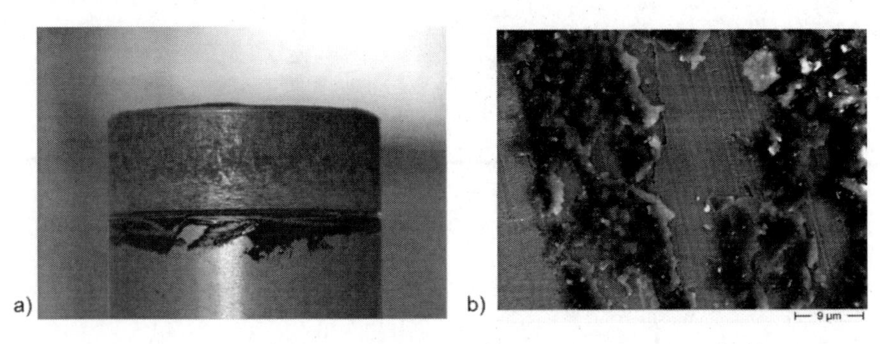

Figure 5. a) Image and b) SEM magnification of a shaft of a hybrid joint at 35 μm of interference and bonded with the *FT-EP* adhesive after the decoupling phase.

The fracture surfaces of hubs and shafts were observed and a section of the bonded area of the shaft for each sample was analyzed at the Scanning Electron Microscope (SEM). Very few differences were observed between samples bonded with the *FT-EP* at different interference levels: it was difficult to detect a clear adhesive or cohesive failure modes as residues of the adhesive layer were present on both the mating surfaces.

The decoupling curves of the hybrid samples prepared with the *FT-EP* adhesive were characterized by two main phases: an initial load peak and a following additional load with a stick/slip trend. The initial peak was due to the contribution of the adhesive plus the interference. The adhesive break up point occurred in correspondence with the maximum of the peak. After the adhesive failure, in order to complete the decoupling, an additional load is necessary to exceed the interference contribution.

The hybrid joint samples with three different classes of interference (10, 22,5 and 35 μm) provided three different additional loads after the break up of the adhesive. Some examples of load-displacement curves related to the decoupling of hybrid joints at different interference levels and of an adhesive joint in clearance condition are shown in figure 6 (Gallio et al., 2014).

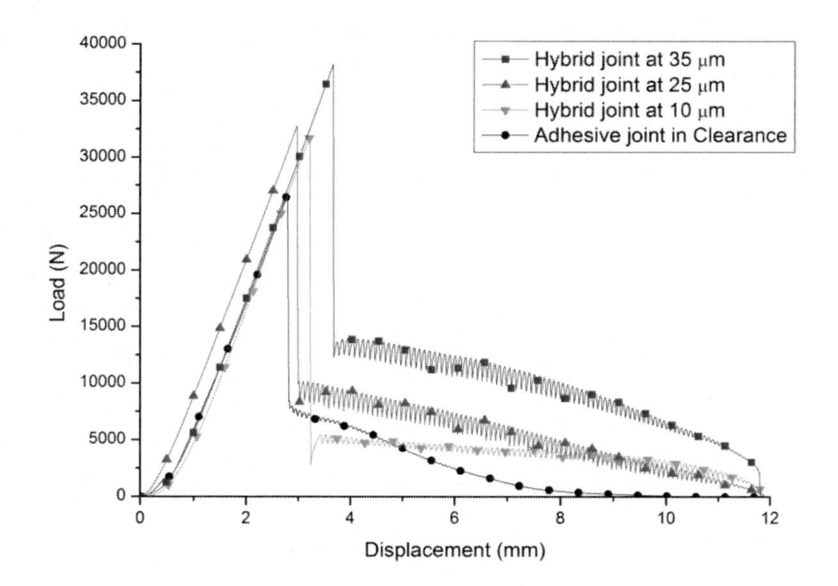

Figure 6. Decoupling curves of three hybrid joints at different interferences and an adhesive joint in clearance condition. At least 4 samples for each class of interference were tested, only one sample is shown as an example in the graph (with permission of Gallio et al., 2014).

The adhesive joints in clearance conditions were also characterized by a residual load level after the break up of the adhesive, due to the friction forces between the cured adhesive residues present on the fracture surfaces. This load presented a different behavior with respect to the interference residual load recorded in the hybrid joints, since it expired before the complete decoupling of the joint.

The load after break in the hybrid joints mainly depended on the radial contact pressure present between the hub and the shaft due to the interference fit. The important contribution due to the friction and wear phenomena observed in the unbonded samples was absent in these joints, thus the load after break can be most likely referred to the system described by the analytical equations. According to this hypothesis, the load after break can be more easily related to the contribution provided by the interference level. In order to obtain an estimation of this additional load, the mean value between the first peak and first valley of the stick and slip state was calculated. The obtained values are plotted in figure 7 and compared to those of the unbonded interference samples.

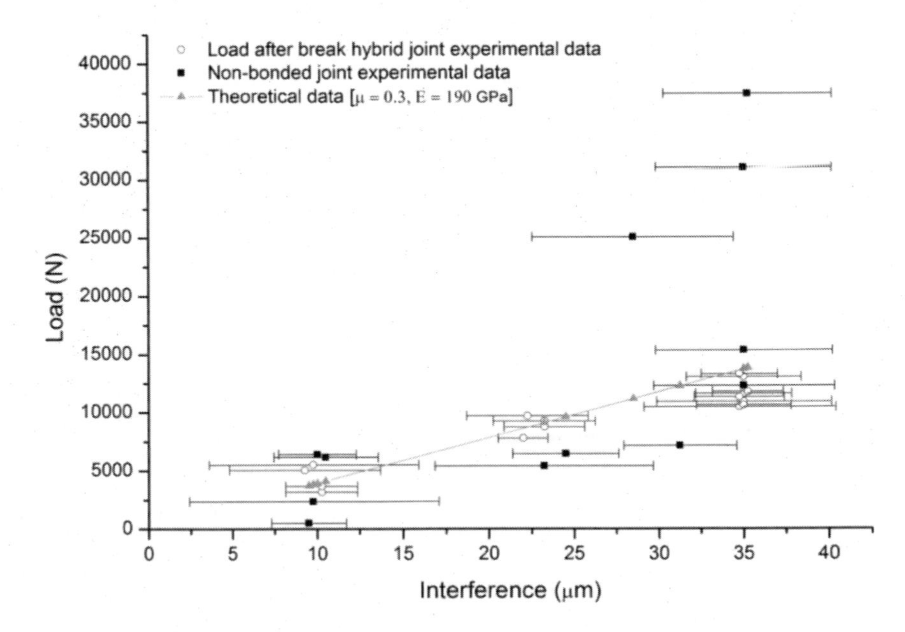

Figure 7. Comparison among the interference contribution detected from the load after break of hybrid joints, the unbonded interference samples, and the results obtained from the analytical equations.

As highlighted in figure 7, the loads after break of the hybrid joints were in good agreement with the theoretical results of F_{int} obtained by processing the measured dimensions of the samples with the Lamè thick walled cylindrical theory and the equation 1, employing a coefficient of friction equal to 0.3 (simulating slightly lubricated surfaces) and a Modulus of the steel adherends of 190 GPa. This relation confirms the assumption of the load after break of hybrid joints as the interference contribution in these samples (Gallio et al., 2014).

Stiffness of the Assembly

The presence of the interference could also affect the stiffness of the hybrid system. The first elastic stretch of the curves of the hybrid joints were compared with the ones of the systems bonded with clearance, considering all the different adhesives tested. The results are reported in figure 8 (Gallio et al., 2013).

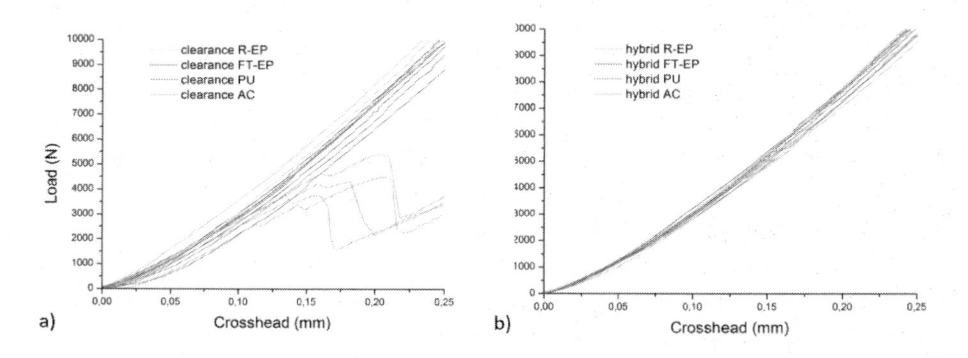

Figure 8. First elastic stretch of: a) the samples bonded with clearance; b) the interference-fitted-adhesive bonded samples (with permission of Gallio et al., 2013).

The differences in the slopes of the curves recorded by testing the bonded samples with clearance were moderated in the hybrid systems by the interference. Thus, it seemed that the interference could be able to control the stiffness of the assembly.

In clearance conditions, among the examined adhesives, the *R-EP* provided the stiffest samples, while the more flexible *PU* adhesive provided the samples with the lowest stiffness. The *FT-EP* samples were characterized by intermediate values. Unfortunately, no precise and comparable information about the Young Modulus of the employed adhesives can be found in their data-sheets. Among the two component adhesives we can say that the epoxy *R-EP* is a rigid adhesive with a Young Modulus of 6700 N/mm^2 and an elongation at break of 0,8 %. The *FT-EP* is a more flexible and toughened blend characterized by a high peel resistance of 92 N/cm, but no comparable data are available. Finally, the *PU* adhesive is characterized by a lower Modulus (576 N/mm^2) and a higher elongation at break (39 %).

The Fatigue Behaviour

A tension–tension fatigue test was performed on the hybrid joints bonded with the *FT-EP* at 35 μm of interference and the *FT-EP* bonded samples with clearance, in order to investigate the influence of the interference on the fatigue behavior of the joints.

The same load amplitude was employed both for the hybrid samples and the bonded-with-clearance ones. The static resistances used as reference for both the cases were calculated from the previous experiments (Gallio et al., 2014).

The results obtained for the hybrid joints were not comparable with the clearance adhesive samples in terms of absolute values of the amplitude load. In fact, the hybrid joints presented higher statical resistance (of about 7000 N) than the clearance ones, so they were expected to begin to suffer the effects of fatigue cycles at higher amplitudes. Nevertheless, it was possible to compare the two cases in terms of a percentage of their static resistance.

The comparison of the fatigue behavior of the hybrid and clearance joints is depicted in figure 9. It could be noted that the hybrid joints underwent the total amount of the fatigue cycles up to a amplitude of about 85 % of their static strength. For higher amplitudes, the samples did not pass the test, failing at about half million of cycles at a maximum amplitude of 90 % and at the beginning of the test for an amplitude amounting to 95 %. On the other hand, the adhesive joints in clearance condition failed around 10^5 cycles if stressed with a max amplitude equal to only 80 % of their static strength, and underwent the million cycles at 60 %. It seemed that the interference played a beneficial role on the fatigue behavior of the samples, as the hybrid joints managed to sustain 10^6 cycles at amplitudes equal to higher percentages of their static resistance with respect to the clearance joints. Both the systems were characterized by similar resonance frequencies of 198.5 ± 1.5 Hz.

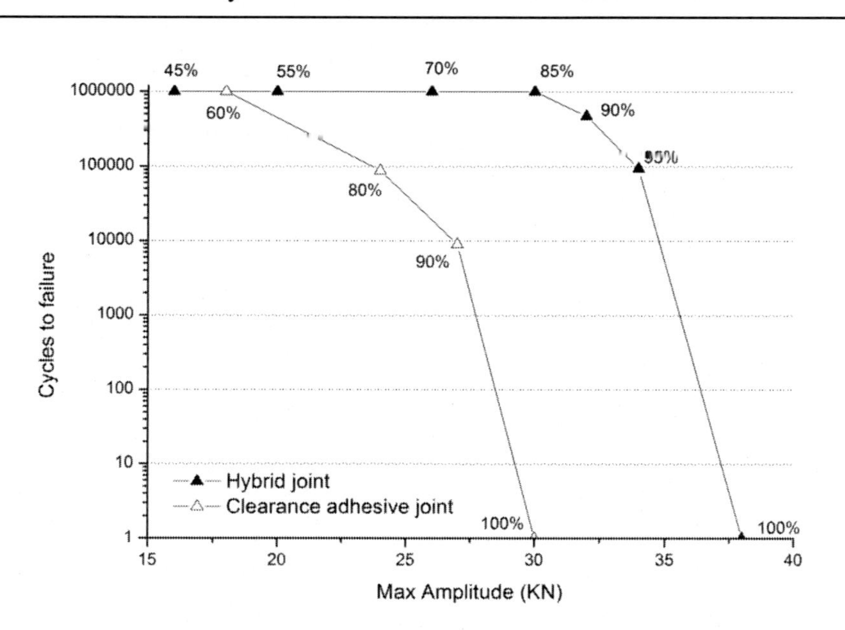

Figure 9. Comparison of the attained cycles to failure of hybrid joints and clearance adhesive joints at difference maximum stress amplitudes. The F_{max} in percentage of their total static resistance is indicated for both case.

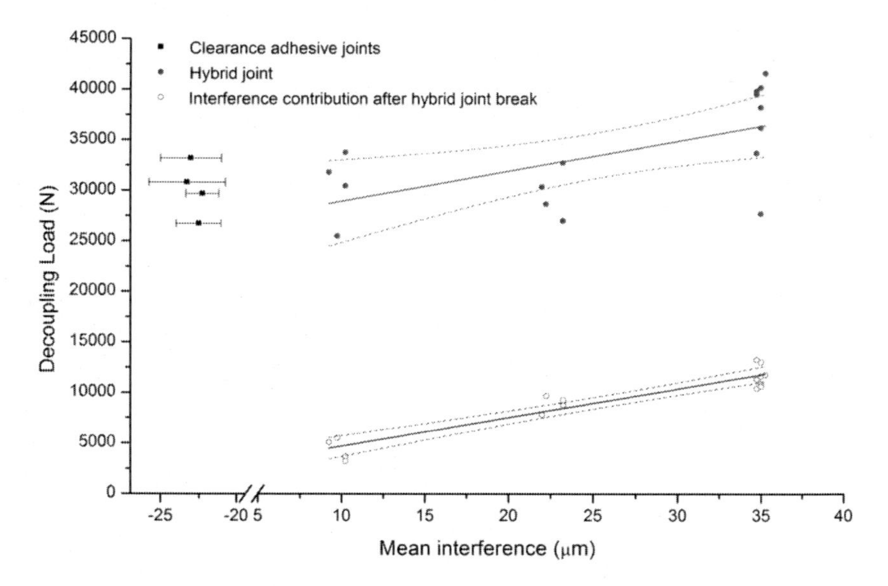

Figure 10. The pull-out loads of the hybrid joints, the interference contribution (detected by the load after break of the hybrid joints) and the adhesive joints in clearance condition as a function of the interference level.

The Adhesive Contribution

In order to understand the importance of the adhesive contribution the total resistances of the hybrid joints bonded with the *FT-EP* at different interference levels were compared to the interference contributions detected from the loads after break of the same samples (figure 10). In addition the resistances of adhesive joints in clearance condition are also included.

The interference enhanced the joint strength of the hybrid systems, especially for higher interference levels. The two linear fits related to the hybrid joints and the interference contributions were characterized by similar slopes, suggesting that the resistance of the hybrid joint is enhanced accordingly to the interference contribution. This parallelism can be referred to the theory of the superposition of the effects of the pressure between hub and shaft and the resistance of the adhesive. But, the superposition theory cannot be absolutely confirmed because of the high dispersion of the data of the hybrid joints resistance. This variability was probably introduced by the adhesive contribution, since the load recorded after the break of the same hybrid joints was clearly dependent on the interference level, with a low standard deviation. Thus, the contribution of the adhesive seemed to govern the total strength of the hybrid joint to a large extent. As a matter of fact, the strength values of the adhesive joints in clearance condition were in the same order of magnitude (Gallio et al., 2014).

Different Adhesive Behaviour

The importance of the adhesive contribution was more clear by the experimental investigation carried out on different adhesive systems. The mean adhesion strength values of the hybrid and clearance joints bonded with different adhesives were extrapolated and the related data are reported in figure 11 (Gallio et al., 2013).

From figure 11 it was possible to observe that the different adhesives presented a diverse behavior in the presence of the same interference level. Indeed, a similar interference contribution for every hybrid system could be assumed, taking into account a mean interference value of 17 ± 4 μm among the specimens. The resultant total strength of the hybrid joints changed on the basis of the adhesive type.

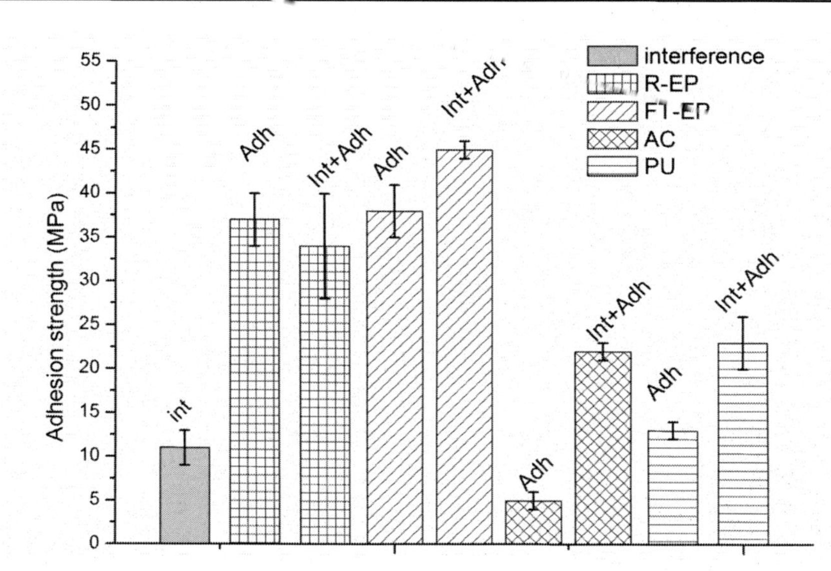

Figure 11. Comparison between the adhesion strength values of all the hybrid and clearance joints bonded with different adhesives. The interference contribution is labeled as "Int", the bonded with clearance systems as "Adh", the hybrid systems as "Int+Adh" (with permission of Gallio et al., 2013).

The hybrid joints prepared with the *FT-EP, AC* and *PU* adhesives presented an increase in the adhesion strength with respect to the unbonded samples. The use of the *FT-EP* adhesive provided values of maximum decoupling load four times higher than those of the interference alone. In the case of the *PU* adhesive, the interference decoupling load was doubled in the hybrid joints. The *AC* adhesive presented the highest resistance improvement in the presence of the interference-fit compared to its performances in clearance condition because of its anaerobic curing mechanism (Gallio et al., 2013).

The *R-EP* was the only adhesive that seemed to be negatively affected by the interference. This result was surely affected by the high data dispersion obtained with the samples bonded with this adhesive. This recorded dispersion was probably due to the quantity of this adhesive able to remain inside the joint during the press-fit coupling. This hypothesis was confirmed by the analysis of the fracture surfaces of every hub and shaft after the decoupling. In the majority of cases, the shafts and the hubs presented some cured adhesive residues on the mating surfaces. In the case of *FT-EP*, *AC* and *PU* samples, the cured residues were present in a similar morphology on both the hubs and the shafts. On the contrary, different residue quantities of *R-EP* remained on the shafts of the various samples (Gallio et al., 2013).

The spreading of the adhesive in interference fit joints depends on the flow properties of the adhesive that are studied by rheology. For what concern the rheology of the adhesives tested in this work, the *FT-EP*, *AC* and *PU* were classified as *thixotropic* (3M Company, n.d.; Henkel Ag & Co. KGaA., 2011; Huntsman Corporation, 2013), whereas this characteristic was not declared from the supplier in the case of the *R-EP* adhesive (Henkel Ag & Co. KGaA., 2011). Thixotropic adhesives are capable of reversible solid-liquid transition: they are *non-sag* pastes that lower their viscosity when subjected to mechanical stresses (Petrie, 2007). In a rest condition and under low shear forces, their particles form a network structure that breaks down when a critical shear rate is reached (Fredrickson, 1964; Gupta, 2000). Generally a low viscosity is recommended in closer joints, while a high viscosity is preferable when a gap filling is required (Henkel Ag & Co. KGaA., 2011). During the press-fit coupling, considerable stresses were developed at the interface, therefore a thixotropic behavior could be advantageous. Probably the *R-EP* adhesive could be spewed away of the joint in a greater extent with respect to the other, thixotropic adhesives. Further studies have to be carried out to deeper analyze the relation between adhesive rheology and its residual quantity in the junction.

APPLICATION OF THE HYBRID JOINT TO THE AUTOMOTIVE WHEEL

Nowadays steel wheels and aluminum alloy wheels own about 50 % of the market share. Often aluminum alloy wheels are preferred for their thin-spoke appearance, but in a field dominated by the constant request of weight and consumption reduction, the automakers more frequently opt for steel wheels to trim the weight of their products as they are lighter and much more cost effective than their aluminum equivalent (American Iron and Steel Institute, 2004; Crain Communications, 2013; M.W. Italia SpA, n.d.).

The current steel wheel is composed of two components, the disc and the rim, joined together by force fitting one into the other and then welded, as schematized in figure 12. Welding is a classic method to join steel components together, but often it is not possible to weld two different metals or joining metals to ceramic, composite or polymeric materials. The exploitation of new materials and new metal alloys is a key solution for obtaining lightweight structures and more innovative and attractive products, with the consequent requirement of a revision of the joining technology.

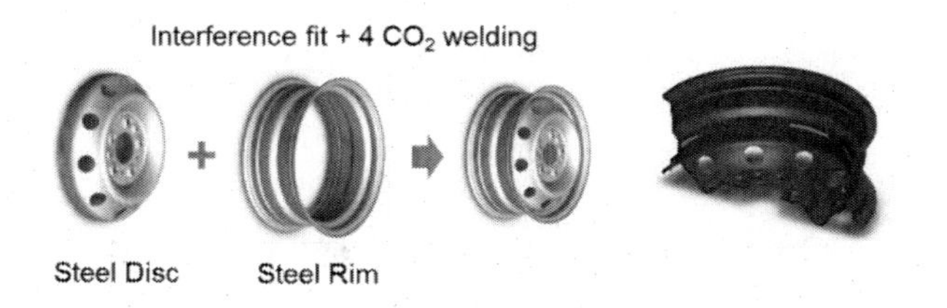

Figure 12.Schematic representation of the present joining technology of the wheel system.

The presence of the interference fit in the wheel is necessary for structural purposes. The introduction of an adhesive instead of the welding in the interference-fit system can lead to an interference fitted-adhesive bonded hybrid joint in the wheel. The wheel has a relatively complex design, where many variables can affect its structural behavior. For instance, the disc flange is not straight, but it is characterized by an inclination. The extent of this inclination is indicated by the *flange angle*, f_α, and generates different interference levels: a lower interference at the level of the upper flange diameter, *UFD*; a stronger interference at the level of the lower flange diameter, *LFD*, as depicted in figure 13. The not univocal interference level plus the plastic deformations of the rim during the press-fit generate a radial contact pressure along the joint area that is not constant and difficult to estimate.

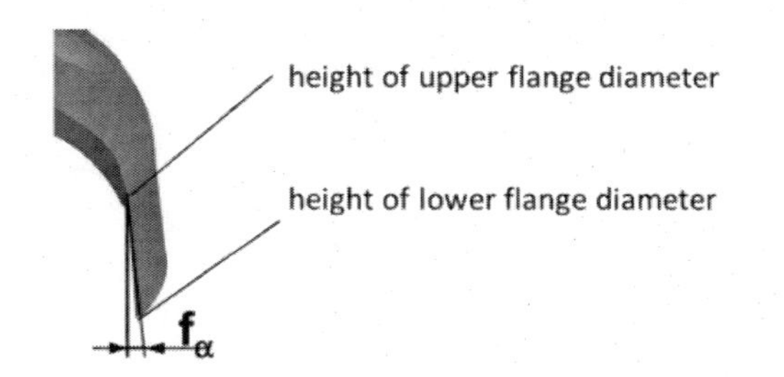

Figure 13. Schematic representation of the disc flangle inclination identified by the flange angle, f_α.

Preliminary static tests were carried out on bonded wheel prototypes in order to understand their general structural resistance and to evaluate the interaction between the adhesive and the interference level in the wheel system. A passenger car wheel was used. In order to create various levels of fitting force, rims were produced with different diameters. The normal production discs were coupled with the different rims obtaining three separate interference classes: stronger interference (A), lower interference (B), transition fit (C). All the interferences and diameters of the sample classes are collected in table 3.

Table 3. Diameters and interferences of the A, B and C classes

Interference classes	Interference at the *UFD* level (mm)	Interference at the *LFD* level (mm)
A	≈ 1.3	≈ 3.1
B	≈ 0.1	≈ 1.9
C	≈ -0.5	≈ 1.3

Wheel Coupling

The coupling load/displacement curves of the bonded wheels were collected during press-fitting. The mean curves of five samples for each case (A, B and C) are reported in figure 14.

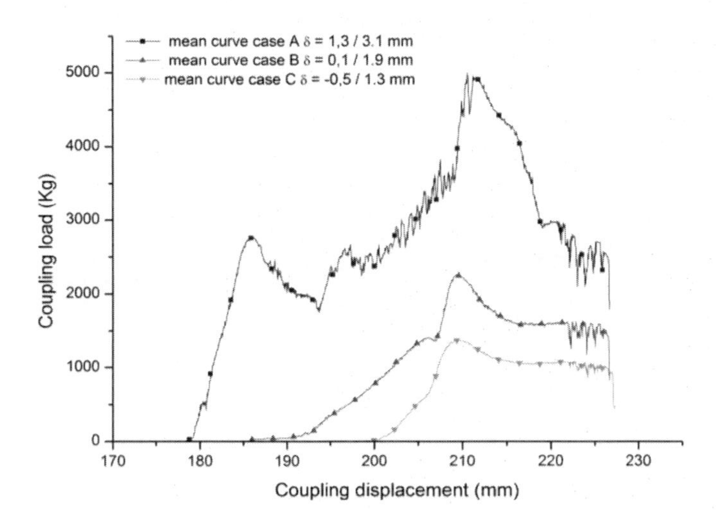

Figure 14. Comparison among the mean coupling curves of bonded wheels for the different fitting-force cases.

All the samples were assembled by using the same discs having the same flange angle. Thus, two interference levels were calculated for the *UFD* and the *LFD*, as reported in table 3. The peak of the entrance of the thinner upper part of the disc flange was visible at 185 mm only for the case A. The peak was very smooth in the case B, corresponding to an interference level at the *UFD* practically inexistent (0.1 mm). For what concern the case C, a clearance was present between the upper part of the flange and the rim well, thus no load was recorded at this displacement. On the other hand, the peak at 210 mm was due to the entrance in the rim well of the larger end of the disc flange. Indeed, it increased from case C to case A according to the interference levels at the *LFD*.

Shear Test and Fractographic Analysis

The wheel *shear test* provides precise information on the shear resistance of the wheel joint. In this test, the disk is pulled out from the rim that cannot be plastically deformed, because of constraints. Also the disk is not able to deform as the load acts on a disk shaped mold placed inside it. Thus, in this way, all the plastic deformations are avoided and all the stresses are concentrated in the joint area, as can be seen in the schematic representation of figure 15.

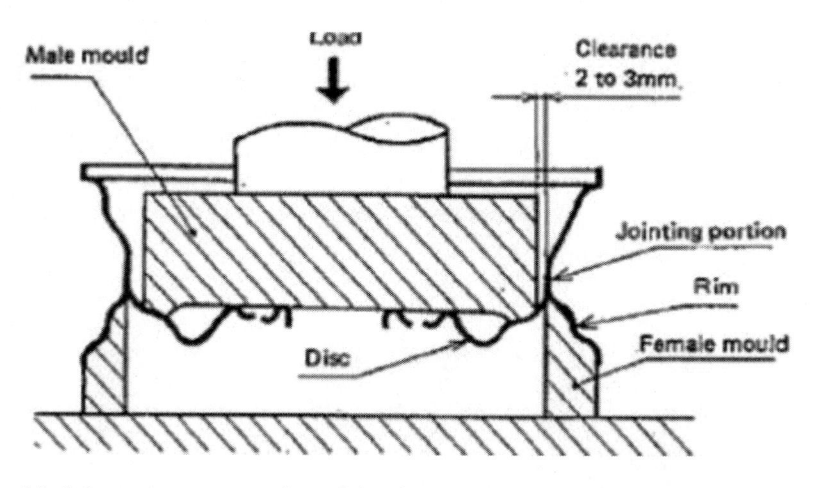

Figure 15. Schematic representation of the shear test on wheel (M.W. Italia SpA, n.d.).

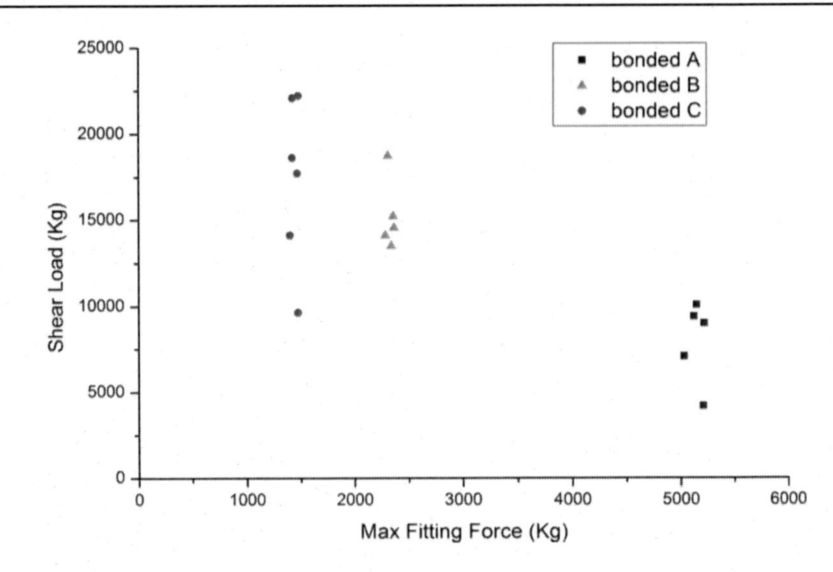

Figure 16. Shear test results of bonded wheels joined with different interference levels. Attained shear loads against the fitting force recorded during coupling operations.

A minimum of five bonded wheels for each case (A, B and C) were submitted to the shear test. On the basis of the laboratory tests on the hub/haft samples, the increase in the interference should imply an increment in the shear strength. On the contrary, the shear tests on bonded wheels revealed an opposite trend, as illustrated in figure 16 where the shear strength is plotted against the maximum fitting force. The shear decoupling load of the case C (the transition fit) was the highest; on the contrary, the case A, characterized by the highest interference, showed the poorest results.

The reason of these unexpected results was explainable through the fractographic analysis of the wheel bonded areas. Discs and rims were cut and analyzed at the optical microscope. A typical fracture surface for each case study is illustrated in figure 17.

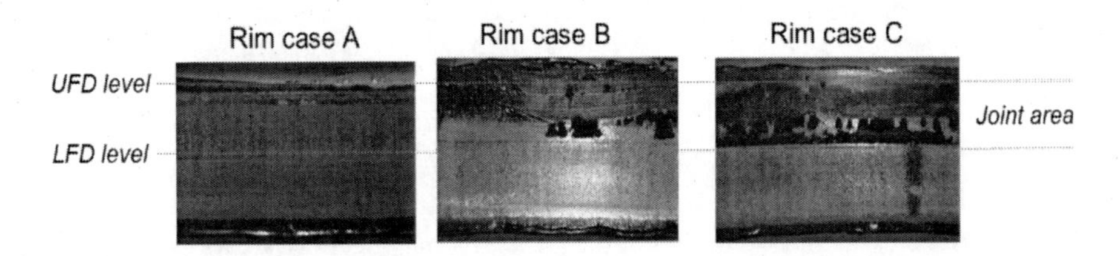

Figure 17. Rim fracture surfaces for each case study after the shear test. The joint area is underlined between the UFD and LFD diameters.

When the disc was coupled into the rim, the adhesive deposited on both the surfaces was dragged. The inclination of the disc combined with the enlargement of the rim due to plastic deformations created a clearance inside the junction. On the contrary, the extremities of the joint area were characterized by a higher interference.

An higher *UFD* interference level limited the adhesive penetration inside the joint. A very limited contact area involving both the adhesive and the interference was present in the case A. The adhesive residues showed a clear fracture surface not in the whole mating area, but only in restricted circumferential stripes at the height of *UFD* and *LFD*. A different situation was observed in the samples B that were characterized by a wider fracture surface in correspondence of the *LFD*. A relevant quantity of adhesive was able to flow inside the junction during the press-fit thank to the lower interference at the *UFD*. The adhesive was then accumulated at the end of the joint gathered by the *LFD* interference. The non-contact area was present also in these samples, but with reduced dimensions with respect to the case A. In the case C, thanks to the clearance at the *UFD* level the adhesive gathered by the interference at the *LFD* level was more than in the case B.

The higher loads reached in shear in the case C seemed plausible considering the quantity of glue that actually formed the bonded joint inside the mating surfaces. The adhesive contribution could be considered higher in the samples B and C than in the case A. It is then possible to resume that samples with higher interference levels showed less shear strength values because they were characterized by a considerably lower adhesive contribution.

Fatigue Behaviour

The *rolling test* was employed to simulate the fatigue stress that the wheel undergoes during its operating life on vehicles. In this test the wheel is placed on a roller and it is subjected to rolling cycles. The machine is able to apply a radial load and a lateral load to simulate the use on vehicles, with a constant monitoring of these parameters during the test. The loading scheme of the test is depicted in figure 18.

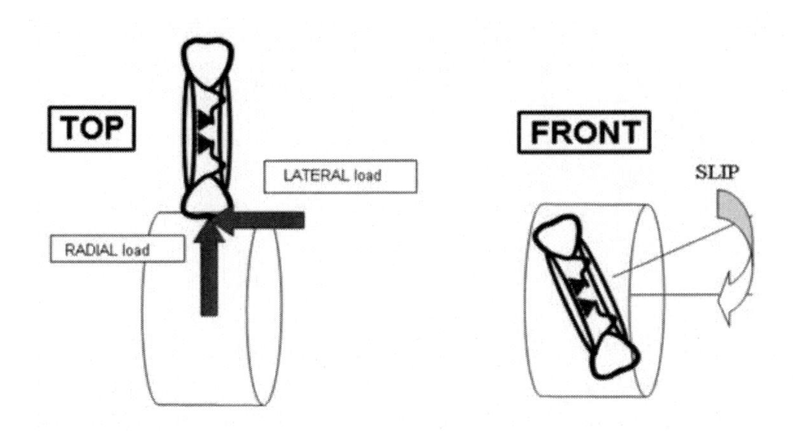

Figure 18. The loading scheme of the rolling test on wheels (M.W. Italia SpA, n.d.).

The fatigue cycle parameters were chosen on the basis of the most harsh standard fatigue test for the examined passenger car wheel. According to the quality protocol (M.W. Italia SpA, n.d.), the 90 % of the normal production wheel should undergo a minimum number of cycles equal to 330000. Both the bonded wheel prototypes and the "traditional" welded wheel extensively overcome the minimum requirement, reaching more than 1.5 millions of cycles without breaking. Considering the long times of test (two or three days for each one), in order to obtain comparable data in a reasonable amount of time, it was decided to further increase the test loads of 15 %. The radial load was then fixed to 1173 Kg and the lateral one to 470 Kg. According to these new values, the speed of the test was fixed at 30 Km/h for safety purpose in the case of decoupling of disc and rim, and in order to reduce the wear of the tires.

Two bonded and two welded wheels for the cases A, B and C were tested in these conditions. The obtained results are resumed in figure 19.

Notwithstanding the limited number of fatigue tests conducted, the obtained results showed a peculiar situation. It seemed that both the systems reached an apex of their fatigue resistance in the case of the samples B (low interference level). The wheels characterized with higher interference levels (case A) and transition fits (case C) failed at lower number of cycles. Moreover, meanwhile the bonded wheels provided repeatable data, a high variability was recorded for the welded wheels, probably imputable to the quality of the welding.

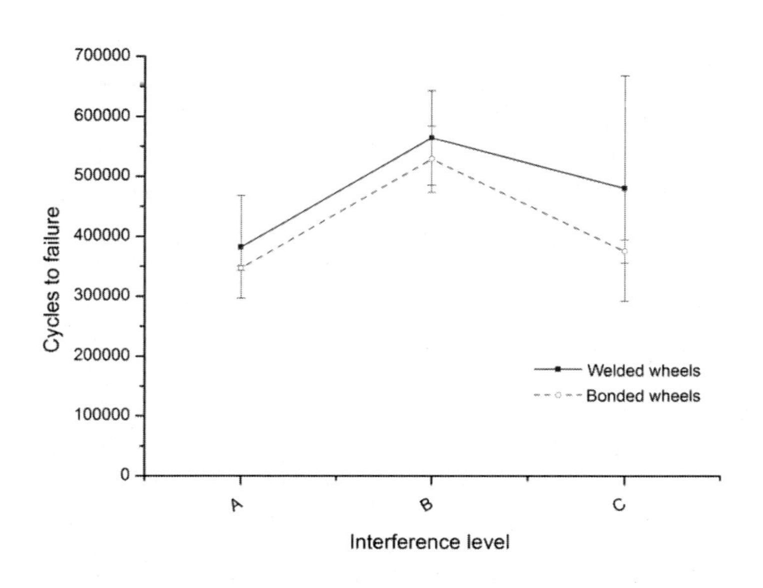

Figure 19. Results of the rolling tests of A, B and C wheels with the radial and lateral loads augmented of 15%.

It was reasonable to state that the shear resistance of the bonded wheel and its fatigue behavior were not correlated. In fact, the case C showed the best performances in static shear but a worse fatigue behavior. While the size of the bonded area seemed to govern the decoupling resistance, the interference seemed to play the biggest role in ruling the fatigue behavior. Indeed, the interference level can be related to parameters such as the vibration frequency and the stiffness, which are important in the field of fatigue analysis.

CONCLUSION

The resultant strength of the examined hybrid interference fitted-ahdesive bonded joints was not always a result of the sum between the adhesive resistance and the interference contribution. The experiments carried out at laboratory level showed that different adhesive types provided dissimilar behaviors in the hybrid joint. Indeed, a greater influence of the adhesive on the maximum strength attainable for the hybrid joint was observed. Rheology of adhesive played a relevant role in determining how much adhesive was spewed away from the joint. If a few quantity of adhesive remained inside the joint the adhesive contribution resulted lowered, also affecting the total resistance of the hybrid joint. In the case of the flexibilized and thoughened epoxy adhesive, the enhancement provided by the interference contribution seemed to influence in the same way the total resistance of the hybrid joints.

The exploitation of adhesives with different flexibility did not provide changes in the stiffness of the hybrid systems, which seemed rather dependent on the interference. Moreover, the presence of the interference positively affected the fatigue behavior of the cylindrical joints. Hybrid joints sustained 10^6 fatigue cycles at load amplitudes equal to an higher percentage of their static resistance compared to the clearance joints.

In the application of the hybrid joint in the wheel system, it was observed that the joint geometry played a relevant role on its performances. The presence of the disc flange angle and the deformability of the rim well created zones of clearance inside the joint. The more the interference the less the quantity of adhesive penetrated inside the clearance. The extent of the bonded area strongly affected the total static resistance of the hybrid joint. On the contrary, the fatigue life of the bonded wheel seemed to be independent to its resistance to the decoupling, and to depend on the interference level to a large extent.

REFERENCES

Abenojar, J., Ballesteros, Y., del Real, J. C., & Martinez, M. A. (2013). Pin-andcollar test method. In *Testing Adhesive Joints: Best Practices* (pp. 155–160). Wiley-VCH.

Adams, R. D., Comyn, J., & Wake, W. C. (1997). *Structural Adhesive Joints in Engineering*. London, UK: Chapman & Hall.

Amancio-Filho, S. T., & Santos, J. F. (2009). Joining of Polymers and Polymer – Metal Hybrid Structures: *Recent Developments and Trends. Polymer Engineering and Science,* 49(8), 1461–1476.

American Iron and Steel Institute. (2004). *Steelworks, the online resource for steel.*

Astm D1002-01. (2001). *Standard test method for apparent shear strength of single-lap-joint adhesively bonded metal specimens by tension loading (metal-to-metal).*

Astm D5656-01. (2001). *Standard test method for thick-adherend metal lapshear joints for determination of the stress-strain behavior of adhesives in shear by tension loading.*

Bowden, F. P., & Tabor, D. (1950). *The friction and lubrication of solids.*

Castagnetti, D., & Dragoni, E. (2012). Predicting the macroscopic shear strength of adhesively-bonded friction interfaces by microscale finite element simulations. *Computational Materials Science,* 64, 146–150.

Castagnetti, D., & Dragoni, E. (2013). Experimental Assessment of a Micro-Mechanical Model for the Static Strength of Hybrid Friction-Bonded Interfaces. *The Journal of Adhesion,* 89, 642–659.

Comyn, J. (1997). *Adhesion Science.* London, UK: The Royal Society of Chemistry.

Crain Communications, I. (2013). *Automotive news.*

Croccolo, D., De Agostinis, M., & Mauri, P. (2013). Influence of the assembly process on the shear strength of shaft–hub hybrid joints. International *Journal of Adhesion and Adhesives*, 44, 174–179.

Croccolo, D., De Agostinis, M., & Vincenzi, N. (2010). Static and dynamic strength evaluation of interference fit and adhesively bonded cylindrical joints. *International Journal of Adhesion and Adhesives,* 30(5), 359–366.

Croccolo, D., de Agostinis, M., & Vincenzi, N. (2011). Experimental Analysis of Static and Fatigue Strength Properties in Press-Fitted and Adhesively Bonded Steel–Aluminium Components. *Journal of Adhesion Science and Technology,* 25(18), 2521–2538.

Da Silva, L., Pirondi, A., & Ochsner, A. (2011). *Hybrid Adhesive Joints.* (L. da Silva, A. Pirondi, & A. Ochsner, Eds.). Heidelberg: Springer.

Dillard, A. (2010). *Advances in structural adhesive bonding* (Dillard A.). New york: CRC Press.

Dragoni, E. (2003). Fatigue testing of taper press fits bonded with anaerobic adhesives. *The Journal of Adhesion,* 79(8-9), 729–747.

Dragoni, E., & Mauri, P. (2000). Intrinsic static strength of friction interfaces augmented with anaerobic adhesives. *International Journal of Adhesion and Adhesives,* 20(4), 315–321.

Dragoni, E., & Mauri, P. (2002). Cumulative static strength of tightened joints bonded with anaerobic adhesives. Proceedings of the Institution of Mechanical Engineers, Part L: *Journal of Materials Design and Applications,* 216(1), 9–15.

Fredrickson, A. G. (1964). Principles and applications of rheology. New York: Prentice-Hall.

Gallio, G., Lombardi, M., Rovarino, D., Fino, P., & Montanaro, L. (2013). Influence of the mechanical behaviour of different adhesives on an interference-fit cylindrical joint. *International Journal of Adhesion and Adhesives,* 47, 63–68.

Gallio, G., Marcuccio, G., Bonisoli, E., Tornincasa, S., Pezzini, D., Ugues, D., … Montanaro, L. (2014). Study of the interference contribution on the performance of an adhesive bonded press-fitted cylindrical joint. *International Journal of Adhesion and Adhesives.*

Gupta, R. K. (2000). *Polymer and composite rheology.* Marcel Dekker Inc.

Henkel Ag & Co. KGaA. (2011). *Henkel Product Selector,* issue 2.

Huntsman Corporation. (2013). *AdMat Araldite 2000plus selector guide.*

ISO 286-1:2010. (2010). *Geometrical product specifications (GPS) - ISO code system for tolerances on linear sizes.*

Kawamura, H., Sawa, T., Yoneno, M., & Nakamura, T. (2003). Effect of fitted position on stress distribution and strength of a bonded shrink fitted joint subjected to torsion. *International Journal of Adhesion and Adhesives,* 23(2), 131–140.

Kinloch, A. J. (1990). *Adhesion and Adhesives.* London, UK: Chapman & Hall.

Kleiner, F., & Fleischmann, W. (2011). Technologies of Threadlocking and Interference-Fit Adhesive Joints. In L. da Silva, A. Pirondi, & A. Ochsner (Eds.), *Hybrid Adhesive Joints* (p. 227). Heidelberg: Springer.

M.W. Italia SpA. *Archive.*

Mengel, R., Häberle, J., & Schlimmer, M. (2007). Mechanical properties of hub/shaft joints adhesively bonded and cured under hydrostatic pressure. *International Journal of Adhesion and Adhesives,* 27(7), 568–573.

Moane, S., Raftery, D. P., Smyth, M. R., & Leonard, R. G. (1999). Decomposition of peroxides by transition metal ions in anaerobic adhesive cure chemistry. *International Journal of Adhesion and Adhesives,* 19(1), 49–57.

O'Reilly, C. (1990). Designing bonded cylindrical joints for automotive applications. In *SAE International Congress and Exposition* (p. paper 900776).

Oinonen, a., & Marquis, G. (2011). A parametric shear damage evolution model for combined clamped and adhesively bonded interfaces. *Engineering Fracture Mechanics,* 78(1), 163–174.

Oinonen, A., & Marquis, G. (2011). International Journal of Adhesion & Adhesives Shear decohesion of clamped abraded steel interfaces reinforced with epoxy adhesive. *International Journal of Adhesion and Adhesives,* 31(6), 550–558.

Petrie, E. M. (2007). *Handbook of Adhesive and Sealants.* New york: Mc Graw Hill.

Sawa, T., Yoneno, M., & Motegi, Y. (2001). Stress analysis and strength evaluation of bonded shrink fitted joints subjected to torsional loads. *Journal of Adhesion Science and Technology,* 15(1), 23–42.

Sekercioglu, T. (2005). Shear strength estimation of adhesively bonded cylindrical components under static loading using the genetic algorithm approach. *International Journal of Adhesion and Adhesives,* 25(4), 352–357.

Sekercioglu, T., Gulsoz, A., & Rende, H. (2005). The effects of bonding clearance and interference fit on the strength of adhesively bonded cylindrical components. *Materials & Design,* 26(4), 377–381.

Yoneno, M., Sawa, T., Shimotakahara, K., & Motegi, Y. (1997). Axisymmetric Stress Analysis and Strength of Bonded Shrink-Fitted Joints Subjected to Push-Off Forces. *JSME International Journal Series A,* 40(4), 362–374.

In: Adhesives
Editor: Dario Croccolo

ISBN: 978-1-63117-653-1

Chapter 3

EFFECT OF THE ENGAGEMENT RATIO ON THE SHEAR STRENGTH AND DECOUPLING RESISTANCE OF HYBRID JOINTS

D. Croccolo*[*]*, M. De Agostinis, G. Olmi and P. Mauri
University of Bologna - DIN, Bologna, Italy

ABSTRACT

Press fitted and adhesively bonded joints give several advantages: among others, they lead to smaller constructions for a given load capacity, or to stronger constructions for a given size. For that reason, the static and fatigue strength properties of these joints have been studied extensively. Studies dealing with the influence of the Engagement Ratio (i. e. the coupling length over the coupling diameter) are still limited. This work aims at filling the gap, limited to the case of a single component anaerobic adhesive. Coupling and decoupling tests have been performed both on press-fitted and adhesively bonded specimens and on pin-collar samples, considering four different levels for the engagement ratio. The study shows that the engagement ratio has a negligible effect on the shear strength of the adhesive and also on the relationship between the decoupling and the coupling forces. Moreover, the obtained results show that a too high interference level in press-fitted and adhesively bonded joints may have a detrimental effect on the adhesive strength.

[*] Correspondence information: dario.croccolo@unibo.it.

Keywords: Anaerobic Adhesives, Engagement Ratio, Hybrid-Joints, Slip-Fitted Joints, Shear Strength

ABBREVIATIONS

A	Coupling surface [mm^2]
$D_{ext,H}$	External hub diameter [mm]
$D_{int,H}$	Internal hub diameter [mm]
D_C	Coupling diameter [mm]
$D_{ext,S}$	External shaft diameter [mm]
E	Young's modulus of the shaft and hub material [MPa]
F_{int}	Interference contribution to the axial release force of the joint [N]
F_{ad}	Adhesive contribution to the axial release force of the joint [N]
F_{coupl}	Pushing in force during shaft-hub coupling (peak value) [N]
F_{tot}	Pushing out (or release) force during shaft-hub decoupling (peak value) [N]
K_1	Ratio between the maximum release force and the maximum pushing in load
L_C	Coupling length [mm]
p_C	Coupling pressure [MPa]
Q_H	Ratio between the internal and external hub diameters
R^2	Linear Correlation coefficient for linear regressions
$R_{a,H}$	Arithmetic mean surface roughness, considering the internal surface of the hub [μm]
$R_{a,H}$	Arithmetic mean surface roughness, considering the external surface of the shaft [μm]
U	Nominal interference [mm]
Z	Actual interference [mm]
ε_θ	Strain measured (by a strain gage) on the external surface of the hub in the tangential direction
μ_A	Axial static coefficient of friction
ν	Poisson's ratio
τ_{ad}	Adhesive static shear strength [MPa]

List of Acronyms

ANOVA	Analysis of Variance
ER	Engagement Ratio
HJ	Hybrid Joint (i.e.: press-fitted and adhesively bonded joint)
PFJ	Dry Press-Fitted Joint (i.e.: coupled with mere interference joint)
SFJ	Slip-Fitted Joint (i.e.: Pin-Collar joint)

INTRODUCTION

The research in the field of joint design is nowadays more and more focused on efficient solutions, which are able to ensure the required connection safety and at the same time the reduction of overall structure weight. A possible option consists in interference joints, where the required interfacing pressure and friction are generated by a proper choice of the coupling tolerances. However, achieving strict tolerances usually implies an increase of manufacturing cost. Moreover, an additional drawback of a friction connection between a shaft and a hub consists in the generation of a not negligible tensile stress state in the hub, especially when the interference level is high. For this reason, the adoption of press-fitted and adhesively bonded joints, usually regarded as Hybrid Joints (*HJs*), is getting more and more frequent in industrial engineering, especially when the transmission of great powers and torques is required. The addition of a suitable adhesive makes it possible to reduce the interference level at the interface, by taking advantage of the bonding layer strength. As a consequence, coarse tolerances may be used, with good outcomes from the point of view of manufacturing costs. Moreover, the tensile stress acting on the hub is strongly decreased. More traditional joining techniques (e.g., keys, pins, interference-fits, bolted joints…) may also be replaced by adhesives, otherwise they can be used in combination. As a matter of fact, dry press-fitted couplings (without adhesive) achieve only 20-30% of nominal contact surface, in the case of a metal to metal joint. On the other hand, the adhesive is able to fill the micro spaces between the crests of surface roughness, so that the contact is extended over the entire area of the mating surfaces [1-5]. A possible drawback of bonded joints stands in the strength of *HJs* being quite difficult to estimate, since it depends on several factors, such as the coupling pressure [6], the type of materials in contact [1, 7], the surface roughness [8], the curing type and the curing methodology [9], the operating temperature [10] and the loading type

[11]. The study in [12] deals with the effect of the type and the way of assembling on the resistance of interference fitted and adhesively bonded joints. In particular, the shear strength of the adhesive is compared, considering press-fitted, shrunk-fitted and cryogenic-fitted bonded specimens. For this purpose, pin-collar specimens were manufactured according to ISO 10123 [13] and involved in release tests after press-fit coupling, whereas, bigger sized specimens were used for shrink-fit and cryogenic-fit, so that the thermal dimensional variations could be more significant. The outcomes of that research indicate that the sample dimensions may affect the adhesive response. In particular, the release strength could depend on the engagement ratio ($ER=L_C/D_C$), namely the ratio between the axial length L_C of the shaft-hub sample and the coupling diameter D_C. In the case of press-fitted and adhesively bonded joints it is also possible to relate the maximum coupling force, namely the ultimate force measured during the pushing in operation, to the maximum pushing out load. In particular, in [12, 14], it was retrieved that the two peak forces are directly proportional and that the ratio between the release and the coupling maximum loads is about 1.2 for an *ER* equal to 1. This ratio will be referenced as K_1: as a further point of discussion, it is important to investigate if this result is maintained even for different *ER* values: the main motivation arises from some discrepancies observed when the *ER* becomes lower than 1 [12, 14]. To the authors' best knowledge, no results are available in the literature about the influence of the *ER* on the static strength of the joint and on the ratio between the release and the coupling loads. These two issues have been tackled in the present paper, by running tests on press-fitted and adhesively bonded specimens and on pin-collar samples. The statistical tools of Design of Experiment have been applied for the full randomization of the test orders and for the comparisons of the statistical distributions.

Materials and Methods

Specimen Design

As explained in the Introduction Section, the aim of the research was to study the effect of the *ER* on joint performance. Thus, the first step has consisted in the determination of the most suitable range of the *ER* to be investigated. The Standard ISO 10123 [13], specific for pin-collar specimens, suggests a sample geometry corresponding to *ER*=0.9. Based on this remark, it seemed to be reasonable to focus the study on *ERs* varying from 0.4 to 1.7, i.e., from approximately one half to the double of the reference value. For statistical evidence reasons, four levels have been considered in the aforementioned range, adding the intermediate values of 0.8, very close to that recommended by the Standard [13], and 1.3.

The second step consisted in the design of the shaft and hub specimens to be connected by interference and adhesive (press-fitted and adhesively bonded samples, *HJs*). For the sake of the consistency of results, when comparing the responses of *HJs* to those of pin-collar samples, it seemed to be reasonable not to divert so far from the recommended proportioning in [13]. However, since Ref. [13] is originally intended for slip fit, pin-collar joints, some modifications had to be introduced, to obtain interference at the shaft-hub coupling. Therefore, the external diameter of the pin was slightly increased, whereas the internal bushing diameter was reduced. All the other dimensions, such as the external diameter and the axial length, remained unchanged with respect to the values recommended in [13]. Following this strategy prevented any unexpected effect, due to a different proportioning of the specimen.

The following point consisted in managing the four values of *ER* to be obtained. Considering a longitudinal extension of 11.1mm, the resulting *ER* is 0.8, a bit lower than the recommended one, due to the rearranged diametric dimension. The longitudinal length was modified in order to design hubs with different *ERs*: half length was considered for the lowest level (*ER*=0.4), while 50% and 100% increased dimensions were used for the two top values (*ER*=1.3, 1.7). The other dimensions remained unchanged to get a consistency of the results for different *ERs*. The drawings of specimens for the four different levels of *ER* are shown in Figure 1 (a, b, c, d), while the pin is sketched in Figure 1 (e): nominal interference is within the interval from 20μm and 30μm, corresponding to an ISO H7/s6 coupling [15]. For reasons of statistical evidence ten pieces were manufactured for each hub type (40 hubs as a total), along with 40 shafts.

Adhesive Specifications

The paper deals with the response of a high strength, single component adhesive that cures in absence of oxygen (anaerobic). Its commercial name is LOCTITE648®, a specifically developed product for cylindrical joints with or without interference. The typical product performances are reported in [16].

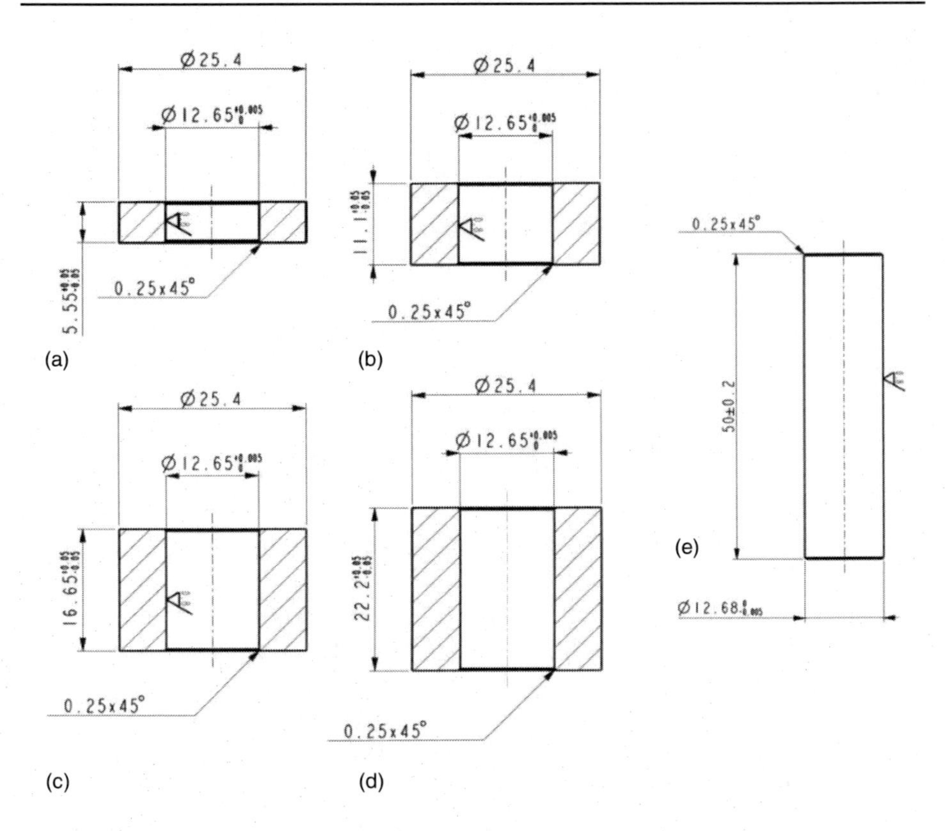

Figure 1. Hub specimens, with *ER*=0.4 (a), *ER*=0.8 (b), *ER*=1.3 (c), *ER*=1.7 (d); shaft sample to be press-fitted into the hubs (e) (all dimensions are in mm).

As exposed above, the effect of *ER* was studied also for adhesively bonded specimens, without interference, namely slip-fitted joints (*SFJs*). For this purpose, pin-and collar samples were properly machined: their proportioning and coupling diameter values comply with [13]. The only dimension to be changed was the overall length of the hubs, in order to retrieve different *ERs*. In this case, only the two extreme levels of the *ER* were considered, corresponding to the lengths of 22mm and 5.5mm. Again, ten pieces per type were manufactured, for a total of 20 pins and 20 collars.

All the shafts and the hubs, both for *HJs* and for *SFJs*, were made of C40 UNI EN 10083-2 steel [17], without any surface treatment. Specimens were machined by means of a CNC lathe, with final grinding treatment in order to reduce surface roughness and to accomplish tolerance requirements.

Theoretical Background

The fundamental design parameter of press-fitted and adhesively bonded couplings is the axial release force F_{tot}. Based on the principle of the superposition of the effects, F_{tot} can be computed as in Eq. (1), by combining the effects of interference (F_{int}) and of adhesion (F_{ad}). The first term, F_{int} [N], depends on the axial static friction coefficient μ_A, on the coupling pressure p_C [MPa] and the contact surface A [mm^2] as indicated in Eq. (2). The second term, F_{ad} [N], is related to the shear strength of the adhesive τ_{ad} [MPa] and, again, to the contact surface A [mm^2] as indicated in Eq. (3). The term A is computed in Eq. (4), where D_C [mm] is the coupling diameter, equal to the mean of the external shaft diameter $D_{ext,S}$ [mm] and the internal hub diameter $D_{int,H}$ [mm], and L_C [mm] is the coupling length.

$$F_{tot} = F_{\text{int}} + F_{ad} \tag{1}$$

$$F_{\text{int}} = \mu_A \cdot p_C \cdot A \tag{2}$$

$$F_{ad} = \tau_{ad} \cdot A \tag{3}$$

$$A = \pi \cdot D_C \cdot L_C \tag{4}$$

It must be pointed out that the principle of superposition of effects (Eq. 1) holds true in the present case, because the adhesive used here belongs to the family of the so called “strong” anaerobics. In the case of “weak” anaerobics (such as LOCTITE243® and other similar ones), more general models must be more conveniently applied, like those put forward by Dragoni and Mauri [18] or by Dragoni and Castagnetti [19].

The axial static friction coefficient μ_A depends on the material characteristics and is strongly influenced by the surface conditions: as a consequence, μ_A may vary in a wide range, between 0.04 and 0.7 according to [20]. Thus, its accurate estimation is important in order to reduce the uncertainty affecting the calculation of F_{int}.

The first step for the determination of the friction coefficient consists in the estimation of the coupling pressure between the shaft and the hub. For this purpose, a strain gage must be applied to the external surface of the hub, along the tangential direction. The relationship between the pressure p_C, the tangential strain ε_θ, and hub material and geometry can be easily retrieved (see Eq. (5)), based on the theory of thick walled cylinders (Lamé's Equation) and on the Hooke's Law.

$$p_C = \frac{E \cdot \varepsilon_\theta \cdot \left(1 - Q_H^2\right)}{2 \cdot Q_H^2} \tag{5}$$

Where Q_H is the ratio between the internal and external diameters of the hub ($Q_H = D_{int,H}/D_{ext,H}$). The practical procedure requires the careful measurement of strain variation, as the shaft is being press fitted into the hub, which is coupled with mere interference, without adhesive (in dry conditions). Upon the completion of the assembly procedure, the coupling pressure is computed by Eq. (5). One of the instrumented hubs is visible in Figure 2 (a), just before starting the pushing in of the shaft. The same hub is depicted after coupling and decoupling in Figures 2 (b) and 2 (c).

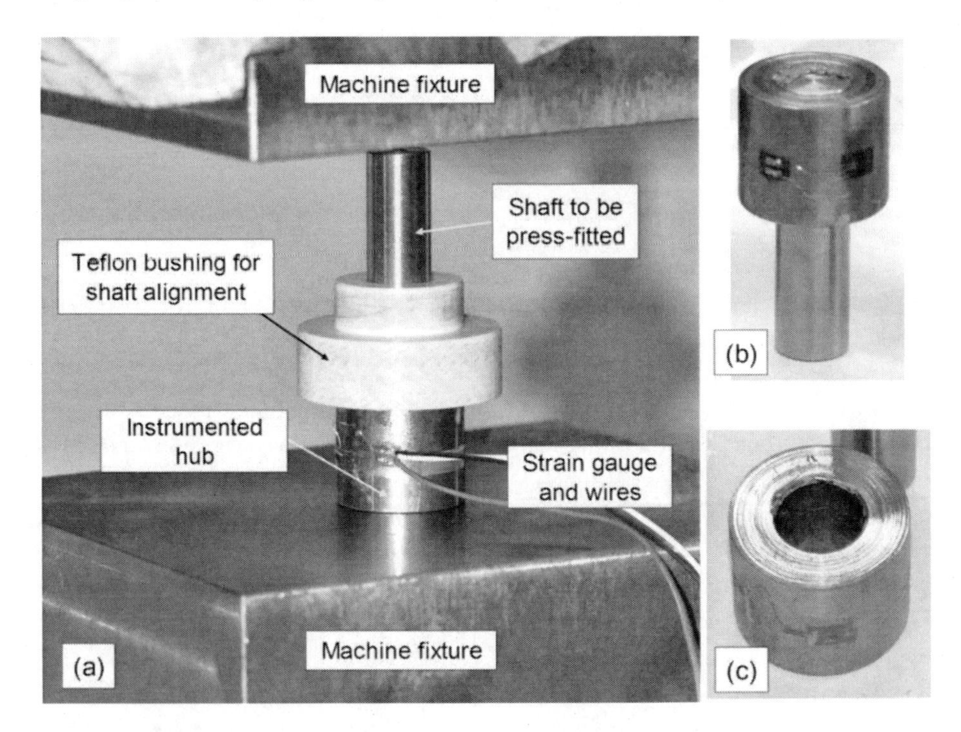

Figure 2. (a) Instrumented hub and experimental setting for shaft insertion, (b) shaft and hub just after pushing in, (c) hub after decoupling.

Once the pressure has been determined, the friction coefficient μ_A can be computed by inverting Eq. (2), as in Eq. (6):

$$\mu_A = \frac{F_{\text{int}}}{p_C \cdot A} \tag{6}$$

The term F_{int} can be determined by running the release test after the shaft and the hub have been assembled under mere interference. The peak of the release force ($F_{int}=F_{tot}$, as $F_{ad}=0$) can be accurately measured by the load cell of the standing press.

In the determination of Fint for each of the press-fitted and adhesively bonded specimens, the coupling pressure $p(Z)$ is determined as a function of the actual interference Z [mm] between the shaft and the hub. The relationship reported in Eq. (7) is valid for shafts and hubs made of the same material.

$$p(Z) = 0.5 \cdot \frac{Z}{D_C} \cdot E \cdot \left(1 - Q_H^2\right) \tag{7}$$

D_C [mm] is the coupling diameter, equal to the mean of the external shaft diameter $D_{ext,S}$ [mm] and the internal hub diameter $D_{int,H}$ [mm], E [MPa] is the Young's modulus. Moreover, the actual interference Z [mm] is lower than the nominal one U [mm], when the shaft is press-fitted into the hub, operating at room temperature. If the stiffness of the parts has a similar value (for example, when they are made of the same material), the difference between U and Z can be related to the surface roughness R_a [mm] of the mating components (hub: $R_{a,H}$; shaft: $R_{a,S}$), as proposed in [14] and reported in Eq. (8).

$$Z = \left(D_{ext,S} - D_{\text{int},H}\right) - 3 \cdot \left(R_{a,H} + R_{a,S}\right) = U - 3 \cdot \left(R_{a,H} + R_{a,S}\right) \tag{8}$$

The reduction of interference in Eq. (8) is related to the crests of surface roughness being chopped during the pin insertion into the hub.

Eq. (1) is recast into Eq. (9), assuming that, in the presence of adhesive, the friction coefficient value μ_A can be considered equal to that relevant to the dry joint. Such occurrence is due to the fact that metals remain in contact in the peaks of the roughness, while the liquid adhesive fills the voids of all depressions [1]. Therefore, F_{ad} can be calculated as the difference between the measured overall strength of the joint (F_{tot}) and the assumed frictional contribution (F_{int}). Finally, by dividing F_{ad} by the mating area, it is possible to estimate the adhesive shear strength τ_{ad}, as indicated in Eq. (10).

$$\begin{aligned} & F_{tot} = F_{\text{int}} + F_{ad} = \mu_A \cdot p_C \cdot A + \tau_{ad} \cdot A \Leftrightarrow \\ & \Leftrightarrow F_{ad} = \tau_{ad} \cdot A = F_{tot} - F_{\text{int}} = F_{tot} - \mu_A \cdot p_C \cdot A \end{aligned} \tag{9}$$

$$\tau_{ad} = \frac{F_{ad}}{A} \tag{10}$$

The term K_1, which was cited in the Introduction Section, can be easily estimated, as in Eq. (11), by monitoring both the coupling and the release force, measured by the loading cell of the standing press, and computing the ratio between the related peak values.

$$K_1 = \frac{F_{tot}}{F_{coupl}} = \frac{F_{\text{int}} + F_{ad}}{F_{coupl}} \tag{11}$$

The determination of τ_{ad} and K_1 makes it possible to discuss the response of the adhesive in the case of *HJs*.

When considering *SFJs*, the coupling procedure is manually performed (as exposed in detail below), so it is not possible to compute the coefficient K_1. The response of the adhesive is represented by the shear stress that can be easily determined, observing that F_{int}=0. Therefore, the ratio between the measured release force F_{tot} and the coupling surface A directly yields τ_{ad}.

EXPERIMENTAL PROCEDURE

The experimentation was split into three campaigns. The first two ones involved *HJs*, with four different *ERs*. Two different interference levels were considered: in particular a higher level in the first campaign and a lower level in the second one. The third campaign involved *SFJs* with two different *ERs* (the extreme levels only were considered).

At a preliminary stage all the shaft and hub specimens to be connected by interference and adhesive were carefully measured by micrometers (for internal and external surfaces) and by a profilometer, in order to determine the actual values of the external diameter of the shaft, of the internal diameter of the hub and of the roughness of the mating surfaces. The samples were arranged in pairs so that the interference values could range in a not too large interval.. This procedure led to a theoretical interference U between 12μm and 26μm, with a mean value of 18μm and a standard deviation of 3μm. The actual interference was estimated by Eq. (8), introducing the measured values for R_a on the mating surfaces. Z stood in the range 8μm-22μm, with an average value of 14μm and a standard deviation of 3μm.

Specimens with different *ERs* were coupled and released, without adhesive, for the estimation of the coupling pressure and of the friction coefficient, as exposed in Section 2. These specimens are later indicated as dry press-fitted joints (*PFJs*)

Before starting the coupling procedure, all the specimens (both shafts and hubs) were accurately cleaned by LOCTITE7063® a multi-purpose cleaner. Then, the adhesive was spread both on the male and on the female parts. Shafts and hubs were connected under a hydraulic press, equipped by a 100kN load cell, operating in the displacement control mode at the fixed actuator velocity of 0.5mm/s. Since misalignments between the mating parts may seriously affect results [21], the coupling procedure was assisted by a specifically developed self-aligning device (partly visible in Figure 2). The coupling force, F_{coupl}, was monitored and sampled throughout each test. Moreover, in order to prevent results from being affected by the test order, this order was fully randomized, involving the entire specimen population, with different *ERs*.

Afterwards, the specimens cured more than 24 hours at room temperature (about 25°C), in order to achieve a complete polymerization [16].

The decoupling procedure was performed by the same standing press and the same fixtures, again with a (different) randomized order of the trials and with on-line monitoring and recording of the exerted release force.

The same specimens were re-used for the second experimental campaign under interference and adhesive. This procedure is justified by the outcomes of previous researches [3, 12], where it was shown that, if only steel components are involved, the specimens can be pushed out and re-used several times without influencing the strength of the joint, provided that they are cleaned accurately after pushing out. For this purpose LOCTITE7063® multi-purpose cleaner was used again.

The male and female parts were measured again for the estimation of coupling dimensions and roughness and then rearranged in (different) pairs. The theoretical interference U was estimated between 2μm and 14μm, with a mean value of 5μm and a standard deviation of 2μm. By subtracting the roughness values, as in Eq. (8), a mean actual interference (*Z*) of 2μm (with a standard deviation of 2μm) was determined. The lower values are due to the general increase of the hub internal diameter, after previous coupling and pushing out.

The same procedure, regarding sample cleaning, order randomizing, applied force recording and specimen curing was carefully followed for both the coupling and the decoupling tests. Two specimens were tested without adhesive for a further estimation of the friction coefficient.

The last stage of the experimental part was devoted to the tests on Pin-and-Collar specimens (*SFJs*). They also were measured, determining the coupling diameters and roughness values. Afterwards they were arranged in pairs, in order to make clearance as uniform as possible over the entire population. The specimens exhibited a clearance between 27μm and 65μm (mean value=48μm; standard deviation=10μm), within the interval 25-75μm recommended by Standard [13], and consistent with the application field of the studied adhesive [16].

Before starting the mating process, all the samples, both shafts and hubs, were cleaned by the aforementioned cleaner. The samples were coupled, applying the adhesive to the external surface of the pin and to the interior of the collar and slipping the collar over the pin, with a helicoidal back-and-forth rapid movement. All the specimens cured for 24 hours at room temperature. The pushing out operation was run with the same procedure followed for *HJs*. The entire specimen population was assembled and disassembled following a randomized order.

RESULTS

As exposed above, three experimental campaigns were performed, involving *HJs* and *SFJs*. Referring to the three campaigns the results are presented in the three sub Sections below. When dealing with *HJs*, the estimation of the coefficient of friction in dry press-fitted conditions is also reported. A comparison between the results of the three campaigns is reported in the Discussion section.

HJs: First Campaign with Higher Interference

In order to estimate the adhesive strength, it was firstly necessary to estimate the coefficient of friction μ_A. The question was tackled by following the procedure in Section 2, involving 5 samples with different *ERs*.The results are reported in Table 1: they can be summarized pointing out that the averaged value for μ_A is 0.22 with a standard deviation of 0.07. After the calculation of the coefficient of friction, the coupling pressure p_C and F_{int} were estimated for each specimen. Following the procedure reported in Section 2, the amount of force that can be attributed to the adhesive, F_{ad}, and its strength, τ_{ad}, were computed. Finally, the coefficient K_1, obtained by the ratio F_{tot}/F_{coupl}, was estimated. All the results, in particular the values of τ_{ad} and K_1, are shown in Table 2.

The results, for τ_{ad} and K_1, are compared in histograms in Figure 3, with indication of the related ranges of variation. Considering all the tests on shaft-hub pairs (independently of the *ER*), the shear strength has a mean value of 17.6MPa and an overall standard deviation of 4.38MPa, while K_1 is on the average around 1.09: these levels are also reported with a dashed line in Figure 3.

The trends of the coupling forces for *PFJs* and *HJs* are compared in Figure 4. The two curves were determined by testing specimens with the same *ER*=0.4. The trends of release loads, obtained when decoupling the same specimens, are sketched in Figure 5. Plotting the pushing in and pushing out curves for all the tested specimens made it possible to determine the peak values, both for F_{coupl} and for F_{tot}, later used for data processing.

Table 1. Summary of the results for *PFJs*, determination of the friction coefficient (first campaign)

PFJs	*ER*	*U* [mm]	*Z* [mm]	p_C [MPa]	$F_{tot}=F_{int}$ [kN]	μ_A
1	0.4	0.019	0.016	75	4.1	0.21
2	0.4	0.021	0.019	84	2.4	0.11
3	0.8	0.019	0.016	46	5.8	0.25
4	0.8	0.021	0.018	56	8.2	0.29
5	1.3	0.021	0.017	40	7.4	0.24
Mean		0.020	0.017	60	5.58	0.22

Table 2. Summary of the results for *HJs* with ER=0.4; 0.8; 1.3; 1.7 (first campaign)

HJs (ER=0.4)	***U* [mm]**	**F_{coupl} [kN]**	**F_{tot} [kN]**	**τ_{ad} [MPa]**	**K_1**
1	0.020	6.3	7.4	13.9	1.18
2	0.019	7.2	8.0	19.0	1.11
3	0.018	7.0	7.8	18.6	1.12
4	0.018	8.2	9.1	25.0	1.11
5	0.014	6.5	7.6	23.5	1.17
6	0.012	6.3	7.0	22.1	1.12
Mean	0.017	6.9	7.8	20.4	1.14
St. Dev.	0.003	0.2	0.7	4.0	0.03
HJs (ER=0.8)	***U* [mm]**	**F_{coupl} [kN]**	**F_{tot} [kN]**	**τ_{ad} [MPa]**	**K_1**
7	0.018	12.3	12.8	10.2	1.03
8	0.016	14.3	16.5	22.5	1.15
9	0.017	15.1	15.2	17.9	1.01
10	0.015	12.5	11.8	11.1	0.94
11	0.012	9.9	12.3	16.7	1.24
12	0.012	11.8	12.8	17.2	1.08
13	0.012	9.3	10.4	13.5	1.12
Mean	0.015	12.2	13.1	15.6	1.08
St. Dev.	0.002	2.1	2.1	4.3	0.10
HJs (ER=1.3)	***U* [mm]**	**F_{coupl} [kN]**	**F_{tot} [kN]**	**τ_{ad} [MPa]**	**K_1**
14	0.020	19.5	22.1	18.2	1.13
15	0.020	25.4	25.9	19.6	1.02
16	0.020	20.4	20.8	15.0	1.02
17	0.020	21.5	24.4	22.2	1.13
18	0.020	22.2	23.7	16.5	1.07
19	0.018	25.4	26.2	26.0	1.03
20	0.018	17.5	18.9	13.6	1.08
Mean	0.019	21.7	23.2	18.7	1.07
St. Dev.	0.001	2.9	2.7	4.3	0.05
HJs (ER=1.7)	***U* [mm]**	**F_{coupl} [kN]**	**F_{tot} [kN]**	**τ_{ad} [MPa]**	**K_1**
21	0.015	25.8	26.9	14.6	1.04
22	0.013	24.7	27.1	17.0	1.10
23	0.013	26.3	29.1	20.4	1.11
24	0.012	19.1	20.7	11.5	1.08
Mean	0.013	24.0	26.0	15.9	1.08
St. Dev.	0.001	3.0	3.7	3.8	0.03

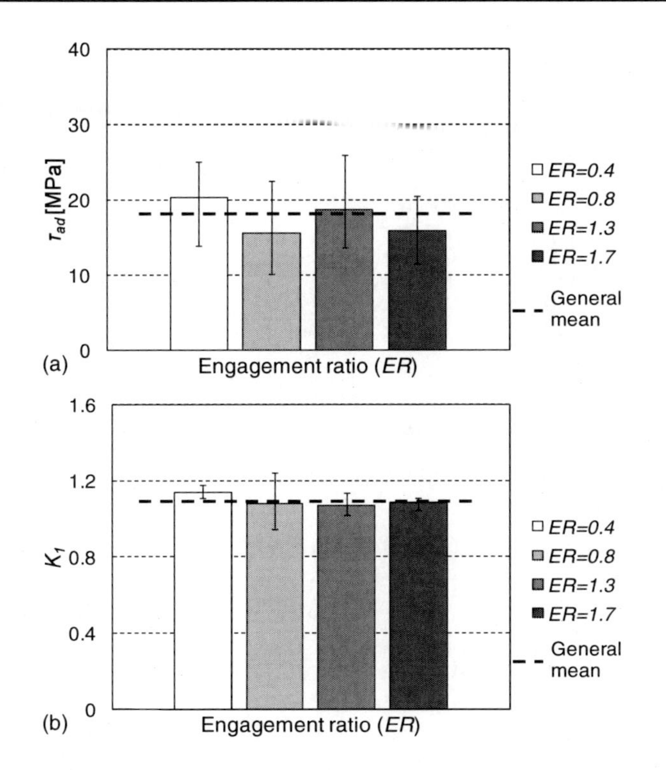

Figure 3. Adhesive shear strengths τad (a) and K1 (b) coefficients compared for different values of *ER* (first campaign on *HJs*).

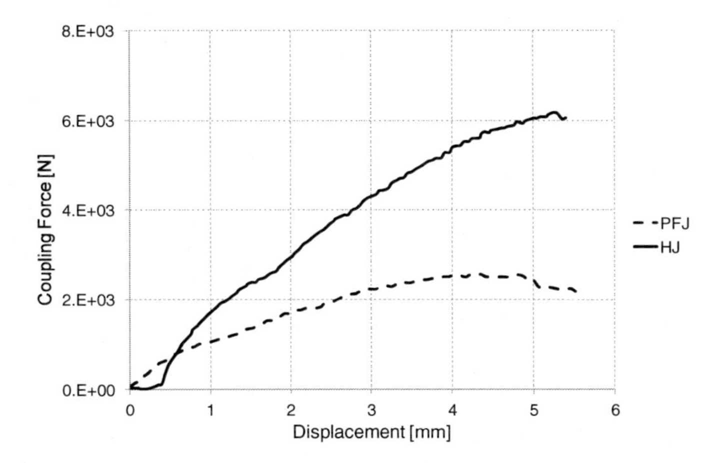

Figure 4. Coupling force of a dry press-fitted joint (*PFJ*) compared to that of a hybrid joint (*HJ*).

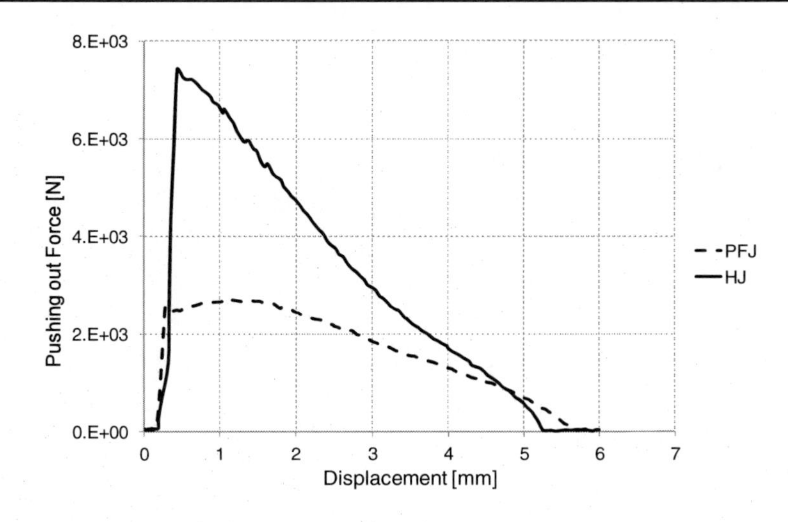

Figure 5. Pushing out force of a dry press-fitted joint (*PFJ*) compared to that of a hybrid joint (*HJ*).

HJs: Second Campaign with Lower Interference

Two specimens (having *ERs* of 0.4 and 0.8) were instrumented by strain gages for the estimation of the coupling pressure and of the coefficient of friction. Detailed results are shown in Table 3: an average coefficient of friction of 0.17 was determined.

All the results, in particular the amount of force sustained by the adhesive, F_{ad}, the strength τ_{ad}, and the ratio K_I, are shown in Table 4. The results, for τ_{ad}, and K_I are compared in histograms in Figure 6, together with their intervals. Considering all the tested shaft-hub pairs (independently of the *ER*), the shear strength has a mean value of 26.2MPa and an overall standard deviation of 5.13MPa, while the K_I ratio is on average around 1.17: these levels also are reported (with a dashed line) in Figure 6.

Table 3. Summary of the results for *PFJs*, determination of the friction coefficient (second campaign)

PFJs	*ER*	*U* [mm]	*Z* [mm]	p_C [MPa]	F_{tot}=F_{int} [kN]	μ_A
1	0.4	0.005	0.002	80	2.7	0.13
2	0.8	0.005	0.002	62	6.5	0.21
Mean		0.005	0.002	71	4.6	0.17

Table 4. Summary of the results for *HJs* with *ER*=0.4; 0.8; 1.3; 1.7 (second campaign)

HJs (ER=0.4)	***U*** **[mm]**	F_{coupl} **[kN]**	F_{tot} **[kN]**	τ_{ad} **[MPa]**	K_1
1	0.008	4.9	4.7	18.5	0.97
2	0.005	6.2	7.4	35.0	1.20
3	0.008	5.2	6.4	31.8	1.23
4	0.009	4.4	5.4	25.7	1.21
5	0.014	6.8	5.8	17.6	0.84
Mean	0.009	5.5	5.9	25.7	1.09
St. Dev.	0.003	1.0	1.0	7.8	0.17
HJs (ER=0.8)	***U*** **[mm]**	F_{coupl} **[kN]**	F_{tot} **[kN]**	τ_{ad} **[MPa]**	K_1
6	0.004	13.0	13.9	31.7	1.07
7	0.004	10.3	10.9	24.0	1.07
8	0.005	9.6	12.8	27.6	1.33
9	0.005	8.7	10.2	21.7	1.17
10	0.008	9.2	12.6	24.3	1.38
11	0.003	11.0	14.3	33.0	1.30
Mean	0.005	10.3	12.5	27.1	1.22
St. Dev.	0.002	1.6	1.6	4.5	0.14
HJs (ER=1.3)	***U*** **[mm]**	F_{coupl} **[kN]**	F_{tot} **[kN]**	τ_{ad} **[MPa]**	K_1
12	0.004	17.1	18.7	30.5	1.09
13	0.004	19.1	18.9	29.7	0.99
14	0.006	17.8	19.3	27.4	1.08
15	0.006	9.4	13.1	18.1	1.39
16	0.003	11.5	12.2	20.1	1.06
17	0.003	16.2	21.4	33.9	1.32
18	0.004	15.7	16.8	24.6	1.07
Mean	0.004	15.3	17.2	26.3	1.10
St. Dev.	0.001	3.5	3.4	5.7	0.15
HJs (ER=1.7)	***U*** **[mm]**	F_{coupl} **[kN]**	F_{tot} **[kN]**	τ_{ad} **[MPa]**	K_1
19	0.004	17.7	19.7	21.6	1.11
20	0.002	23.6	26.3	30.9	1.11
21	0.003	18.0	23.8	26.3	1.32
22	0.003	17.7	20.9	27.1	1.18
23	0.002	16.2	18.7	22.2	1.16
24	0.003	17.1	23.6	26.5	1.38
Mean	0.003	18.4	22.2	25.8	1.21
St. Dev.	0.001	2.6	2.9	3.4	0.11

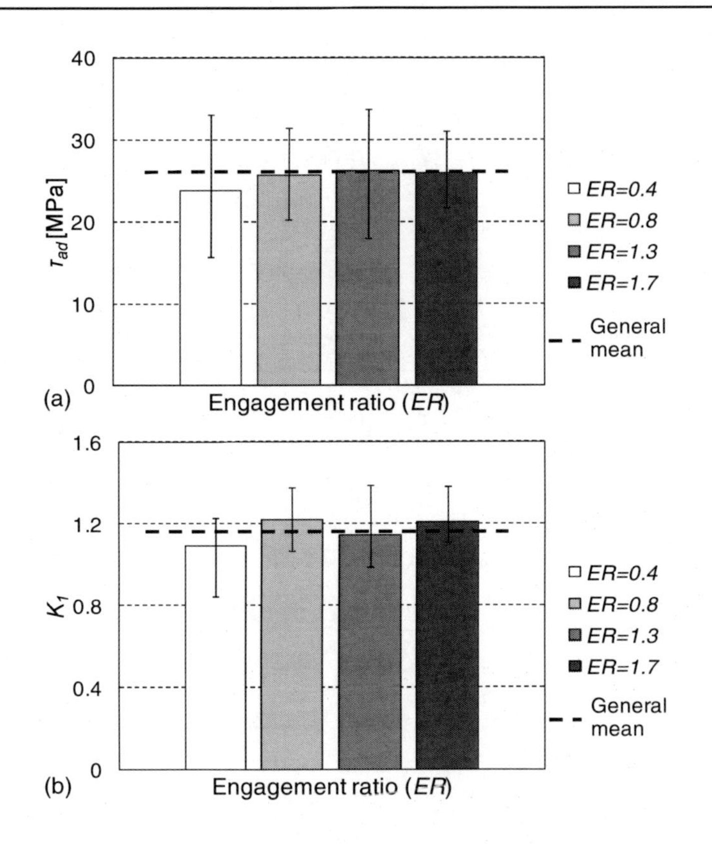

Figure 6. Adhesive shear strengths τ_{ad} (a) and K_I (b) coefficients compared for different values of *ER* (second campaign on *HJs*).

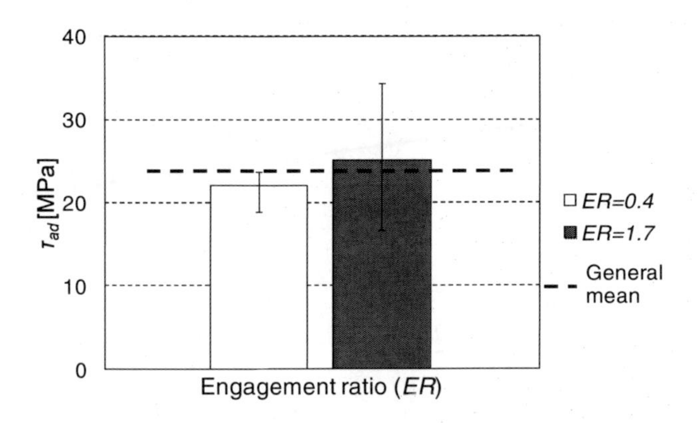

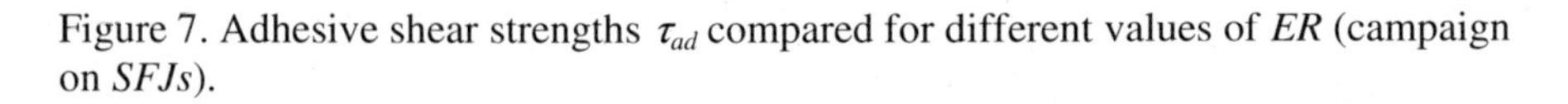
Figure 7. Adhesive shear strengths τ_{ad} compared for different values of *ER* (campaign on *SFJs*).

Determination of the Shear Strength of *SFJs*

The results related to the investigated *ERs*, 0.4 and 1.7, are shown in Table 5. The estimated strengths for *ER*=0.4 and *ER*=1.7 are compared in the histogram in Figure 7, with indication of the mean values and of the variation band. Considering the data altogether, independently of the value of the *ER*, the shear strength is averagely about 23.6MPa and the overall standard deviation is 3.85MPa.

Table 5. Summary of the results for *SFJs* with *ER*=0.4; 1.7

SFJs (ER=0.4)	*Clearance* [mm]	$F_{tot}=F_{ad}$ [kN]	τ_{ad} [MPa]
1	0.057	4.9	22.1
2	0.058	5.2	23.7
3	0.053	4.3	19.5
4	0.058	5.2	23.4
5	0.061	4.2	18.9
6	0.035	5.0	22.7
7	0.061	5.3	23.7
8	0.061	5.0	22.6
Mean	0.056	4.9	22.1
St. Dev.	0.009	0.41	1.86
SFJs (ER=1.7)	*Clearance* [mm]	$F_{tot}=F_{ad}$ [kN]	τ_{ad} [MPa]
9	0.057	20.8	23.5
10	0.050	14.8	16.7
11	0.048	30.4	34.4
12	0.051	22.5	25.4
13	0.057	22.8	25.7
14	0.053	22.4	25.3
15	0.050	22.2	25.1
16	0.052	22.4	25.3
Mean	0.052	22.3	25.2
St. Dev.	0.003	4.23	4.78

DISCUSSION

First of all, by looking at Figures 4-5, observing the curves for *PFJs* joints, it can be remarked that the peak of the release force is often lower than the coupling load maximum value. This result is consistent with previous experimental outcomes [3] and can be explained, observing that the pin insertion implies chopping the roughness crests, with the consequence that the actual interference (Z) is reduced (as explained also in the Materials and Methods section). Therefore, in the release condition the pressure acting on the mating surfaces and the resulting friction forces are lowered. The determined values of pressure are comparable to reference values for steel to steel contact and so are the resulting values of μ_A [3].

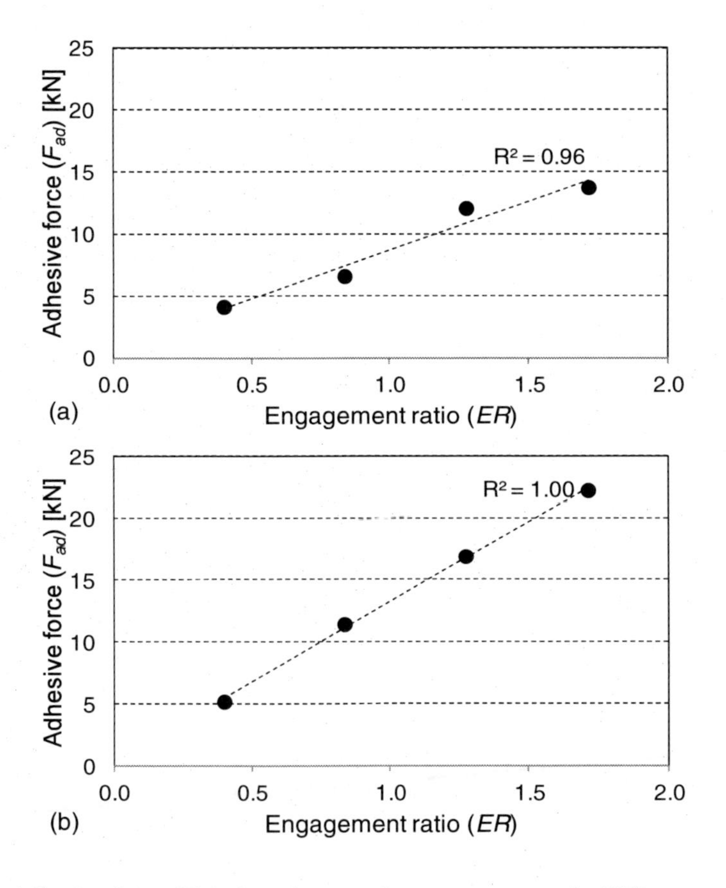

Figure 8. Adhesive force (F_{ad}) plotted versus the engagement ratio (ER), processing the data of the first (a) and of the second (b) campaigns on *HJs*.

The subject of this paper consisted in the investigation of the influence of the *ER* on the adhesive performance. Thus, the main issue of discussion consists in the comparison between the determined values for the shear strengths τ_{ad} and the K_1 ratios. By looking at the histograms in Figures 3 and 6, it can be observed that the values are very close to each other and that the intervals are often overlapped. These data distributions undoubtedly suggest that the *ER* has a poor impact on the variation of the mentioned parameters. Moreover, the data were processed by the statistical tools of Design of Experiment: in particular, the Analysis of Variance (*ANOVA*) and the Fisher test (*F-Test*). The combined use of these tools can be regarded as the most suitable approach to state if a factor has an impact (i.e., a significant effect) on one or more response variables. From the methodological point of view, full details are provided in the Appendix Section, in [22] and in [20, 23-27], where these tools were successfully applied to compare data populations. In the present study the *ANOVA* and the *F-Test* were applied to process the experimental data, in order to establish if the *ER* has an effect on the adhesive strength and on the K_1 ratio. The results of the first and the second campaigns on *HJs* were taken into account, thus investigating the effect of *ER* at different interference levels. All the analyses confirmed that no significant differences can be observed (at the 5% significance level), i.e., the impact of *ER* is negligible. The response of *ER* being not significant is confirmed also by the linear relationship between the amount of force on the adhesive, F_{ad}, and *ER*, considering the results of both experimental campaigns on *HJs*. The related trends are shown in Figure 8: it can be remarked that the linear correlation coefficient (R^2) is almost 1 (indicating perfect linearity) in the second case.

This result can be explained by observing that, assuming τ_{ad} as an invariant, F_{ad} must be proportional to A. By considering Eq. (4), A is undoubtedly proportional to the coupling length L_C, which is proportional to L_C/D_C, i.e., to the engagement ratio. The relationship is summarized in Eq. (12).

$$F_{ad} \propto A \propto L_c \propto \frac{L_C}{D_C} = ER \quad (12)$$

Since it was proved that the *ER* is ineffective, it seems to be correct to consider the averaged values (of τ_{ad} and K_1) over the entire specimen populations, when comparing the adhesive response in the first and in the second experimental campaigns. It was pointed out that the shear strength is about 18MPa considering the first set of data and 26MPa, in the second case, with an increment of almost 50%. The retrieved values must be discussed in the light of the different interference levels: 18μm in the first case and 5μm in the second one (considering the mean values of the theoretical interference *U*).

A possible reason for a worse performance when the interference level is higher is that during the press fit operation a certain amount of adhesive is stripped away from the mating surfaces, thus reducing the actual bonded area. This effect is clearly much more reduced when the interference level is low, or when the coupling procedure is conducted by different techniques, e.g., shrink-fitting [12]. This outcome is consistent with the results in [20], where it was pointed out that a high interference level, expensive for production, is usually not suitable to increase the joint resistance. Photos of a shaft and a hub (*ER*=1.7) after decoupling are shown in Figure 9 (a) with reference to the first campaign, and in Figure 9 (b) with reference to the second. It can be easily observed the adhesive polymerization is much better in the second case, thanks to the lower interference level.

Figure 9. Adhesive polymerization compared in specimens of different sizes with a higher (a, b) and a lower (c, d) interference.

Finally, the ratio K_I is slightly increased (from 1.09 to 1.17), as the interference decreases, indicating a better adhesive response, however it does not encounter strong variations. The retrieved range for K_I is consistent with that in [14]. Therefore, K_I confirms to be a very useful parameter to predict the *HJ* static strength, once the maximum coupling load is known (measured by the load cell of the hydraulic press during the assembly phase). The results of this research confirm that this prediction is reliable, also when the *ER* has values far from 1.

The last remark regards the results of the third campaign, involving pin-collar specimens. No significant differences were also observed between the estimated shear strengths. Moreover, the determined values, around 24 MPa, are in agreement with the expected adhesive performance in pin-collar applications [16].

CONCLUSION

The subject of the present research was to investigate the effect of the ratio between the coupling length and the coupling diameter, namely the engagement ratio, on the response of a high strength single component anaerobic adhesive.

The issue was tackled by manufacturing and testing press-fitted and adhesively bonded joints, namely mixed joints (considering two levels of interference), and pin-collar samples.

The results were processed for the determination of two key design parameters: (i) the adhesive shear strength and (ii) the ratio between the release force and the coupling load. Statistical tests applied to the experimental results led to the conclusion that the engagement ratio has no effect on the joint strength, both with and without interference.

Considering the results of mixed joints, it was found out that the strength is increased from 18 to 26MPa (+50%), as the interference is decreased from 18 to 5 μm (considering the mean values of the theoretical interference *U*). This result is consistent with previous researches and may be explained with reference to the occurrence of the adhesive being partially stripped away during coupling.

Finally, the aforementioned load ratio also proved to be independent of the engagement ratio and to stand in the interval of about 1.1 to 1.2. From the point of view of joint design, this result makes it possible to reliably predict the decoupling force, based on the measured force upon coupling, even when the engagement ratio is far from 1.

APPENDIX: ANALYSIS OF VARIANCE TO COMPARE THE ADHESIVE RESPONSE FOR DIFFERENT ERS

The tool of the one-factor Analysis of Variance (*ANOVA*), combined to the statistical test of Fisher (usually regarded as the *F-test*), is the most suitable methodology to state if a factor has an impact (i.e., a significant effect) on one or more response variables. In the present study the *ANOVA* was applied to process the experimental data, in order to establish if the *ER* has an effect on the adhesive strength and on the K_I ratio. For the sake of clarity the whole procedure is briefly summarized below, before reporting the results concerned with the present research. Full details are contained in Ref. [22] and related references. A preliminary step to the analysis consists in the rearrangement of the data to be processed in a two-dimensional array, where the columns refer to the factor levels, while the rows are related to replicated results. A general example is shown in Table A1, considering C levels for the involved factor and R_j replicated data for each level (the number of replicated data may be different, depending on the level).

Table A1. Sample table of the results to be processed

	Factor Levels					
	Level 1	Level 2	...	Level j	...	Level C
Replications	y_{11}	y_{12}	...	y_{1j}	...	y_{1C}
	y_{21}	y_{22}	...	y_{2j}	...	y_{2C}
	...	...	...	...	...	...
	y_{i1}	y_{i2}	...	y_{ij}	...	y_{iC}
	...	...	...	...	...	...
	$y_{R_1 1}$	...	...	...	...	...
		$y_{R_2 2}$	...	...	...	...
				$y_{R_j j}$	...	...
						$y_{R_C C}$

Let y_{ij} represent the result corresponding to the *(j-th)* level of the factor and to the *(i-th)* replication, and let $\bar{y}_{.j}$ be the average of the data retrieved for the *(j-th)* factor level, listed in the *(j-th)* column. The symbol $\bar{y}_{..}$ indicates the Grand Mean over all the results. The total amount of variance in the experiment, usually regarded as *Total Sum of Squares*, *TSS*, can be computed as in Eq. (A1).

$$TSS = \sum_{j=1}^{C}\left[\sum_{i=1}^{R_j}\left(y_{ij} - \bar{y}_{..}\right)^2\right] \tag{A1}$$

The *TSS* can be split into two terms: the first one accounts for the effect of the studied factor and is called *Sum of Squares Between Columns*, *SSBC*, whereas, the second one takes the experimental uncertainty into account and is regarded as the *Sum of Squares Error*, SSE. The related expressions are shown in Eqs. (A2) and (A3).

$$SSBC = \sum_{j=1}^{C}\left[R_j \cdot \left(\bar{y}_{.j} - \bar{y}_{..}\right)^2\right] \tag{A2}$$

$$SSE = \sum_{j=1}^{C}\left[\sum_{i=1}^{R_j}\left(y_{ij} - \bar{y}_{.j}\right)^2\right] \tag{A3}$$

The question regarding the significance of the factor is tackled, comparing *SSBC* and *SSE*. For this purpose, these two terms must be properly scaled, i.e., divided by the related degrees of freedom, ν_{SSBC} and ν_{SSE} (see Eqs. (A4) and (A5)).

$$\nu_{SSBC} = C - 1 \tag{A4}$$

$$\nu_{SSE} = \sum_{j=1}^{C}\left(R_j - 1\right) \tag{A5}$$

The scaling procedure leads to the computation of the *Mean Squares*, *MSBC* and *MSE*, as in Eqs. (A6) and (A7).

$$MSBC = \frac{SSBC}{\nu_{SSBC}} \tag{A6}$$

$$MSE = \frac{SSE}{\nu_{SSE}} \tag{A7}$$

The following step consists in the computation of the *Fisher ratio*, $F_{calc.}$, i.e., the ratio between *MSBC* and *MSE*.

$$F_{calc.} = \frac{MSBC}{MSE} \tag{A8}$$

Finally, the *p-value*, *p-v.*, is computed as a function of $F_{calc.}$, ν_{SSBC} and ν_{SSE}, considering the *Fisher* distribution of the *Fisher ratio*. This outcome leads to the *F-test*, where the *p-v.* is compared to a significance level α. The final response is that the factor has a significant impact on the output, if p-v.< α. Otherwise, it is concluded that the output is not significantly affected by the factor. The level α is usually assumed as 5%, as suggested in many References, such as [22].

Analyses of variance were performed to compare the results in terms of the adhesive shear strength τ_{ad} and of the ratio K_I for different values of *ERs*, varying in the range from 0.4 to 1.7 (4 levels as a total). The analyses processed both the results of the first campaign on *HJs* (Tables A2-A3) and those retrieved in the second campaign (Tables A4-A5), in order to investigate the effect of *ER* at different interference levels.

Table A2. ANOVA and Fisher test regarding the impact of *ER* on τ_{ad} (first campaign on *HJs*)

Sums of squares	Degrees of Freedom	Mean Squares	$F_{calc.}$	p-v.
SSBC = 95.09	$\nu_{SSBC} = 3$	MSBC = 31.70	1.83	17.4%
SSE = 345.96	$\nu_{SSE} = 20$	MSE = 17.30		
TSS = 441.04				

Table A3. ANOVA and Fisher test regarding the impact of *ER* on K_1 (first campaign on *HJs*)

Sums of squares	Degrees of Freedom	Mean Squares	$F_{calc.}$	p-v.
SSBC = 0.015	$\nu_{SSBC} = 3$	MSBC = 0.005	1.29	30.5%
SSE = 0.080	$\nu_{SSE} = 20$	MSE = 0.004		
TSS = 0.096				

Table A4. ANOVA and Fisher test regarding the impact of *ER* on τ_{ad} (second campaign on *HJs*)

Sums of squares	Degrees of Freedom	Mean Squares	$F_{calc.}$	p-v.
SSBC = 6.77	$\nu_{SSBC} = 3$	MSBC = 2.26	0.08	97.3%
SSE = 598.85	$\nu_{SSE} = 20$	MSE = 29.94		
TSS = 605.63				

Table A5. ANOVA and Fisher test regarding the impact of *ER* on K_1 (second campaign on *HJs*)

Sums of squares	Degrees of Freedom	Mean Squares	$F_{calc.}$	p-v.
SSBC = 0.059	$\nu_{SSBC} = 3$	MSBC = 0.020	0.95	43.4%
SSE = 0.414	$\nu_{SSE} = 20$	MSE = 0.021		
TSS = 0.473				

ACKNOWLEDGMENTS

The Authors would like to gratefully acknowledge Profs. Matteo Mozzini and Stefano Ruggeri and all the students of the "Scuola Specializzata Superiore di Tecnica" in Bellinzona (Switzerland) for their kind support in the manufacturing of shaft and hub specimens. The Authors also acknowledge Eng. Paolo Proli, Director of the DIN Mechanical and Aerospace Lab, for his help in the experimental part of this research.

REFERENCES

[1] Sekercioglu, T., Gulsoz, A., Rende, H., 2005. The effects of bonding clearance and interference fit on the strength of adhesively bonded cylindrical components. *Mater. Design*. 26, 377-381.

[2] Kawamura, H., Sawa, T., Yoneno, M., Nakamura, T., 2003. Effect of fitted position on stress distribution and strength of a bonded shrink fitted joint subjected to torsion. *Intl J. Adhes. Adhes*. 23, 131-140.

[3] Croccolo, D., De Agostinis, M., Vincenzi, N., 2010. Static and dynamic strength evaluation of interference fit and adhesively bonded cylindrical joints. *Intl J. Adhes. Adhes*. 30, 359-366.

[4] Dragoni, E., Mauri, P., 2000. Intrinsic static strength of friction interfaces augmented with anaerobic adhesives. *Intl J. Adhes. Adhes*. 20, 315-321.

[5] Dragoni, E., 2003. Fatigue testing of taper press fits bonded with anaerobic adhesives. *The Journal of Adhesion*. 79, 729-747.

[6] Bartolozzi, G., Croccolo, D., Chiapparini, M., 1999. Research on shaft–hub adhesive and compression coupling. *Öesterr. Ing. Z.*. 5, 198-201.

[7] Sekercioglu, T., 2005. Shear strength estimation of adhesively bonded cylindrical components under static loading using the genetic algorithm approach. *Intl J. Adhes. Adhes*. 25, 352-357.

[8] Sekercioglu, T., Rende, H., Gulsoz, A., Meran, C., 2003. The effects of surface roughness on the strength of adhesively bonded cylindrical components. *J. Mater. Proc. Technol*. 142, 82-86.

[9] Mengel, R., Haberle, J., Schlimmer, M., 2007. Mechanical properties of hub/shaft joints adhesively bonded and cured under hydrostatic pressure. *Intl J. Adhesion Adhesives*. 27, 568-573.

[10] Da Silva, L.F.M., Adams, R.D., 2007. Joint strength predictions for adhesive joints to be used over a wide temperature range. *Intl J. Adhes. Adhes*. 27, 362-379.

[11] Adams, R.D., Comyn, J., Wake, W.C., 1997. Structural Adhesive Joints in Engineering. London, Chapman & Hall.

[12] Croccolo, D., De Agostinis, M., Mauri, P., 2013. Influence of the assembly process on the shear strength of shaft-hub hybrid joints. *Intl J. Adhes. Adhes*. 44, 174-179.

[13] ISO 10123, 1990. *Adhesives - Determination of Shear Strength of Anaerobic Adhesives Using Pin-and-Collar Specimens.*

[14] Croccolo, D., De Agostinis, M., Vincenzi, N., 2011. Experimental analysis of static and fatigue strength properties in press-fitted and

adhesively bonded steel-aluminium components. *Journal of Adhesion Science and Technology*, 25 (18), 2521-2538.

[15] ISO 286-1, 1991. *ISO System of limits and fits-part 1: bases of tolerances, deviations and fits.* Milan: UNI ISO.

[16] Loctite, 2005. *648 Technical Data Sheet.*

[17] UNI EN 10083-2, 2006. Steels for quenching and tempering - Part 2: Technical delivery conditions for non alloy steels.

[18] Dragoni, E., Mauri, P., 2002. Cumulative static strength of tightened joints bonded with anaerobic adhesives. *Proceedings of the Institution of Mechanical Engineers, Part L: J. Materials Design and Applications.* 216, 9-15.

[19] Castagnetti, D., Dragoni, E., 2012. Predicting the macroscopic shear strength of adhesively-bonded friction interfaces by microscale finite element simulations. *Computational Materials Science*. 64, 146-150.

[20] Croccolo, D., Cuppini, R., Vincenzi, N., 2008. Friction coefficient definition in compression–fit couplings applying the DOE method. *Strain*. 44, 170-179.

[21] Olmi, G., 2011. A New Loading-Constraining Device for Mechanical Testing with Misalignment Auto-Compensation. *Exp. Tech.* 35 (6), 61-70.

[22] Berger, P.D., Maurer, R.E., 2002. Experimental Design with applications in management, engineering and the sciences. Belmont, CA, Duxbury Press.

[23] Olmi, G., Freddi, A., Croccolo, D., 2008. In-Field Measurement of Forces and Deformations at the Rear end of a Motorcycle and Structural Optimisation: Experimental-Numerical Approach Aimed at Structural Optimisation. *Strain*. 44, 453-461.

[24] Olmi, G., 2009. Investigation on the influence of temperature variation on the response of miniaturised piezoresistive sensors. *Strain*. 45, 63-76.

[25] Olmi, G., Comandini, M., Freddi, A., 2010. Fatigue on Shot-Peened Gears: Experimentation, Simulation and Sensitivity Analyses. *Strain*. 46, 382-395.

[26] Olmi, G., 2012. Low Cycle Fatigue Experiments on Turbogenerator Steels and a New Method for Defining Confidence Bands. *Journal of Testing and Evaluation*. 40 (4), 539-552.

[27] Croccolo, D., De Agostinis, M., Vincenzi, N., 2011. Failure analysis of bolted joints: effect of friction coefficients in torque-preloading relationship. *Eng. Fail. Anal.* 18, 364-373.

In: Adhesives
Editor: Dario Croccolo

ISBN: 978-1-63117-653-1

Chapter 4

IMPROVEMENT OF PROPERTIES OF HYDRAULIC MORTARS WITH ADDITION OF NANO-TITANIA

N. Maravelaki*[*]*, C. Kapridaki, E. Lionakis and A. Verganelaki
Technical University of Crete, School of Architectural Engineering, Greece

ABSTRACT

In this work nano-titania of anatase form has been added to mortars containing (a): binders of either lime and metakaolin or natural hydraulic lime and, (b): fine aggregates of carbonate nature. The aim was to assess the adhesive performance of the above binders for reassembling fragment porous stones and more specifically to explore the effect of nano-titania in the hydration and carbonation of the derived mortars. The nano-titania proportion was 4.5-6% w/w of binders. The physicochemical and mechanical properties of the nano-titania mortars were studied and compared to the respective ones of the mortars without the nano-titania addition, used as reference. DTA-TG, FTIR, SEM and XRD analyses indicated the evolution of carbonation, hydration and hydraulic compound formation during a period of one-year curing. The mechanical characterization indicated that the mortars with the nano-titania addition showed improved mechanical properties over time, when compared to the

[*] Corresponding Author address: Email: pmaravelaki@isc.tuc.gr.

specimens without nano-titania. The results evidenced carbonation and hydration enhancement of the mortar mixtures with nano-titania. The hydrophylicity of nano-titania enhances humidity retention in mortars, thus facilitating the carbonation and hydration processes. This property can be exploited in the fabrication of mortars for reassembling fragments of porous limestones from monuments, where the presence of humidity controls the mortar setting and adhesion efficiency. A specifically designed mechanical experiment based on direct tensile strength proved the suitability of mortars with nano-titania as a potential adhesive means for restoration applications. The rapid discoloration of methylene blue stains applied to mortars with nano-titania supported the self-cleaning properties of mortars with nano-titania presence. Based on the physico-chemical and mechanical characterization of the studied adhesive mortars with nano-titania, binders of metakaolin-lime and natural hydraulic lime have been selected as most appropriate formulations for the adhesion of fragment porous stones in restoration applications.

Keywords: Adhesive mortars, nano-titania, metakaolin-lime, hydraulic lime, mechanical properties

INTRODUCTION

The effective adhesion of mortars to fragments of archaeological stone or other building materials results in substantial structural integrity and prevents further decay. Treatment options include the application of adhesives and grouts, as well mechanical pinning repairs. Commonly used adhesives, such as epoxy, acrylic and polyester resins and cement mortars demonstrate excessive strength, high irreversibility and, if improperly applied, their removal may be more damaging to the historic fabric (Amstock 2000).

In this research work, two different kinds of stone from Piraeus, namely Aktites and Mounichea stones, corresponding to a hard dolomitic limestone and a marly limestone of the area respectively, were selected due to their common employment as main construction materials for the Athenian Acropolis building during the Archaic period. The preferred treatment strategy for the reassembling and adhesion of these fragments was addressed through designing bonding mortars compatible to these stones. Repair criteria were as follows: (a) physico-chemical and mechanical compatibility between repair materials and stone, (b) adequate strength to resist tensile and shear forces, (c) retreatability, (d) longevity, (e) affordability and (f) ease of installation.

The design of adhesive mortars was based on binders of either hydrate lime-metakaolin or natural hydraulic lime, with the aim of formulating a complex system characterized by the highest compatibility. Nowadays, both hydrate lime-metakaolin and natural hydraulic lime mortars are widely used in the field of restoration and conservation of architectural monuments, due to their capability to enhance the chemical, physical, structural and mechanical compatibility with historical building materials (stones, bricks and mortars) (Rosario 2009). This compatibility is a very critical prerequisite for the optimum performance of conservation mortars, considering the damage caused to historic monuments during the past decades, due to the extensive use of cement-based composites.

In this framework, in the specially designed mortars consisting of binders of either lime and metakaolin or natural hydraulic lime and fine aggregates of carbonate nature, nano-titania of anatase (90 per cent) and rutile (10 per cent) form has been added (4.5-6% w/w of binder). The aim was to study the effect of nano-titania in the hydration and carbonation of the above binders and to compare the physico-chemical properties of the nano-titania mortars with those mortars without nano-titania, used as reference. Thermal analysis (DTA-TG), infrared spectroscopy (FTIR), X-ray diffraction (XRD) and scanning electron microscopy (SEM) analyses were performed to investigate the evolution of carbonation, hydration and hydraulic compound formation during a six-month curing period. Furthermore, the stone–mortar interfaces, the adhesion resistance to external mechanical stress, relative to the physico-chemical characteristics of the stone-mortar system and the role of the nano-titania as additive, were reported and are discussed in this chapter.

METHODS

Design of Adhesive Mortars: Binders, Fillers and Aggregates

Binders of either lime (L: by CaO Hellas) with metakaolin (M: Metastar 501 by Imerys), or natural hydraulic lime (NHL: NHL3.5z by Lafarge), as well as nano-titanium dioxide (T: nano-structured nano-titania by NanoPhos), used as filler due to its photocatalytic activity, were employed for the design of the adhesive mortars. The already established improvement of hydration and carbonation process due to the photocatalytic activity of nano-titania in anatase form (Hyeon-Cheol 2010) added to cement mortars, was taken into consideration to assess whether the adhesion performance of the studied mortars was affected. Moreover, the self-cleaning properties of the adhesive mortars could also be attained due to the photocatalytic action of the nano-titania. XRD, FTIR and DTA-TG techniques were used to characterize the raw products.

Table 1 lists the mortar mixes; the ratio of water to binder (W/B) ranges from 0.8 to 0.6. The quantity of lime required to react with metakaolin was fixed in a weight ratio equal to 1.5, ensuring the pozzolanic reaction. Any unreacted quantity of lime, after its carbonation, provides plasticity to the end-product and enables the mortar to acquire a pore size distribution similar or compatible to porous stone, thus allowing a homogeneous distribution of water and water vapor in the complex system (Moropoulou 2005). Furthermore, this enhanced plasticity can function as a tool for the distribution and absorption of external stresses, which otherwise could lead to the mechanical failure of the mortar. TiO_2 nano-powder dispersion in a small amount of water was achieved through ultrasonic treatment for 15 min; afterwards the obtained TiO_2 colloidal solution was mixed with the other raw materials and the mixture was stirred with a handheld mixer for 5 min. Due to the fact that the fine aggregates can contribute to the avoidance of shrinkage and cracking during the setting process, the addition of sand with fine grains was deemed essential. Consequently, equal proportions of sand passing through the 125 and 63 μm sieves were added in the mix; these were previously washed with water to free any harmful soluble salts.

The above mixtures (Tablc 1) were molded in prismatic and cubic moulds, with dimensions of 4×4×16 cm and 5×5×5 cm, respectively and were then placed in a curing chamber for setting, at RH = 65 ± 3 % and T = 20 ±2 °C, according to the procedure described in the EN 196-1. The mechanical properties of the designed mortars were characterized by measuring their uniaxial compressive (Fc) and flexural (Ff-3pb) strengths, according to EN 1015-11:1999.

Table 1. Mortar mixes (composition in mass %)

Samples	Sand	Binders			Filler	B/A	W/B
Code	Cc	NHL	M	L	Nano-titania (T)		
NHLT1	48	49			3	1	0.7
NHLT2	33	64			3	2	0.6
ML	50		20	30	0	1	0.8
MLT	47		20	30	3	1	0.8

NHL: natural hydraulic lime, M: metakaolin, L: lime, B: binder, A: aggregate, W: water.

Assessment of the Adhesive Mortars

The Aktites and Mounichea stone samples were cut and shaped into specimens with dimensions of: (a) 4×4×4 cm and were used to measure the compressive strength of the stone and (b) 4×4×8 cm that were used to test bonding with the designed mortars. Before applying the adhesive mortars, the stone faces were first incised using special mechanical tools to provide a rough surface and were then wetted. Afterwards, the mortar was partially applied to the stone surface and then the stone specimens were "filled" with mortar by placing them levelly with the aid of special joint clamps (Figure 1b). The joint was kept moist with damp cotton gauze and a polyethylene sheet. The specimens were then placed in a storage chamber for 28 days, according to the conditions described in EN 1015-11:1999. Home designed equipment for both four point flexural (Ff-4pb) (Figure 1a) and direct tensile (Ft) (Figure 1c) strength tests, were used for measuring the adhesive performance of the bonded stone-mortar specimens (4×4×17 cm) (Whittaker 1992).

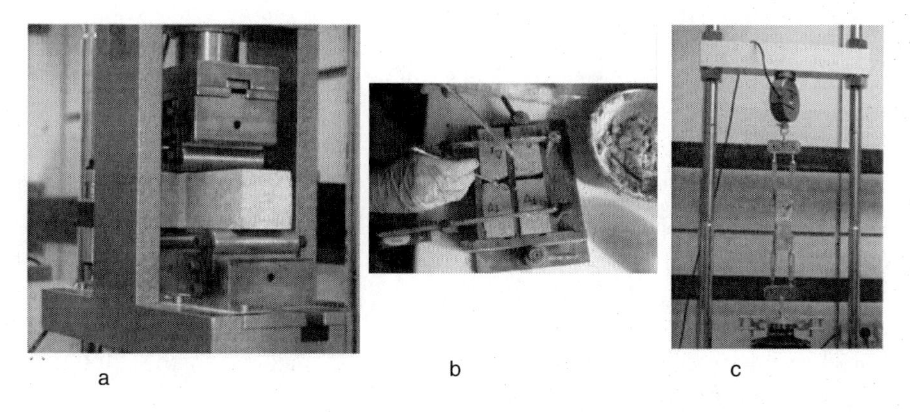

Figure 1. a) Four point bending apparatus with variable support spans b) application of adhesive mortars to stone surfaces c) direct tensile test apparatus for stone mortar specimens.

The evolution of hydration and carbonation of powder samples was assessed by XRD, mercury intrusion porosimetry (MIP), FTIR, DTA/TG, EDXRF and SEM. By identifying calcium silicate hydrates (CSH) and calcium aluminate hydrates (CAH) at different ages, not only qualitatively, but also semi-quantitatively, the hydration and carbonation processes could be monitored. The mineralogical analysis was investigated by XRD with a Siemens D-500 diffractometer (40 kV/35 mA); the spectra were collected between 5° and 60° on the 2θ scale, with a step of 0.03°/5s. SEM analysis was carried out in fractured surfaces, using a FEI Quanta Inspect scanning electron microscope. DTA/TG was operated with a Setaram thermal analyser in static air atmosphere up to 1000 °C at a rate of 10 °C/min. The FTIR analysis of KBr pellets was carried out using a Perkin-Elmer spectrophotometer in the spectral range of 400-4000 cm^{-1}. MIP measurements were recorded using a Quantachrome Autoscan 60 porosimeter, in the range of 2-4000 nm. The physical properties of the stone and mortars were further studied by water absorption at atmospheric pressure, according to EN 13755:2002, as well as by capillary water absorption measurements according to EN 1015-18:1995 (for mortars) and EN 1925:1999 (for stones).

RESULTS AND DISCUSSION

Physico-Chemical Evaluation of Stones to Be Adhered and Adhesive Mortars

The mineralogical composition, according to the Rietveld method, and the measured mechanical strength of stones to be adhered are presented in Table 2. According to these results, the Aktites stone samples (D1, D3) are hard and compact micritic dolomitic limestones, with a low to medium porosity, small grain size and high values of mechanical strength. The Mounichea stone samples are marly limestone/dolomitic limestone with an oolitic texture, brownish or yellowish to light grey-color, with a low to medium porosity and low values of mechanical strength (D2, D4, K). The Aktites samples were characterized by higher compressive strength, compared to the Mounichea samples, as a result of their lower content of clay material.

From the results of the DTA-TG analysis, the total bound water (Htb) identified in the temperature range from 100 to 400 °C corresponds to the dehydration of CSH, CAH and the residual bound water (Aggelakopoulou 2011). The dehydration of $Ca(OH)_2$ (CH) occurred between 480 to 500 °C, while the decomposition of $CaCO_3$ (Cc) took place at a temperature higher than 600 °C.

In particular, Figure 2 depicts the DTA curves for the NHL mortars without nano-titania at 1, 11 and 28 days of curing. The inset plot illustrates the evolution of the ratio unreacted-CH to formed-Cc at the same curing period for NHL mortars with and without nano-titania. As the setting process proceeds, mass losses attributed to the release of CO_2 from $CaCO_3$ increased, while the mass loss of CH dehydration decreased, due to the transformation of CH into hydraulic components and calcite. The lime consumption and the formation of hydraulic phases are more pronounced for the nano-titania mortars at different curing times. The same observation was also obtained from the thermal analyses of metakaolin-lime mortars with and without nano-titania.

Table 2. Mineralogical composition and compressive strength of the stones

Minerals	Calcite (%)	Dolomite (%)	Chlorite (%)	Quartz (%)	Kaolinite (%)	Illite (%)	Albite (%)	Compressive Strength Fc (MPa)
D1	6.2	83.0	0.6	3.7	1.8	4.2	0.5	42.3
D2	1.5	78.7	2.8	6.1	2.5	7.8	0.5	28.7
D3	12.0	80.0	0.9	2.7	2.4	1.9	0.1	107.0
D4	4.2	80.0	1.3	3.3	3.3	7.0	0.9	6.6
K	11.1	70.9	1.1	5.4	0.9	8.1	2.2	22.7

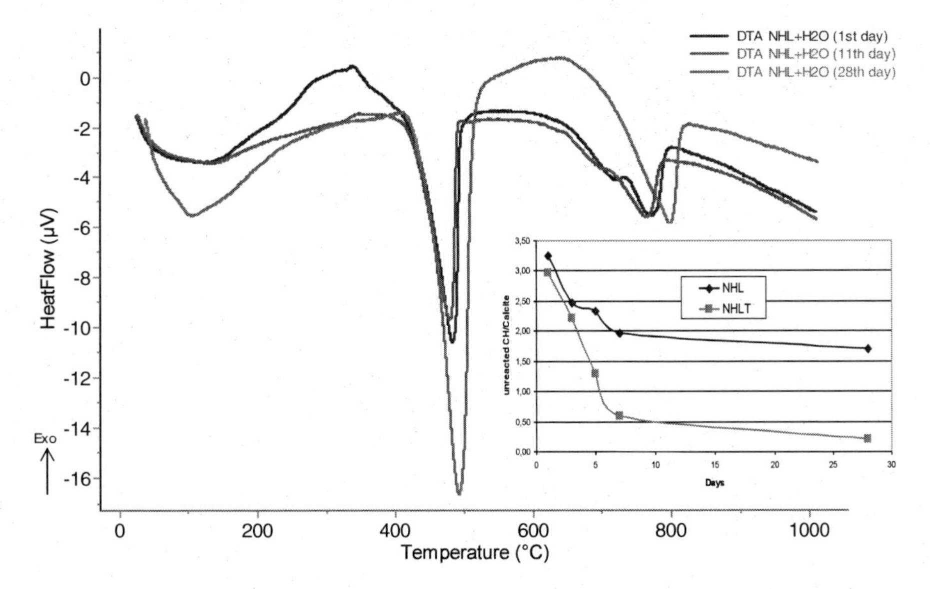

Figure 2. DTA curves for NHL mortars without nano-titania cured for 1, 11 and 28 days along with the evolution of carbonation in NHL mortars without and with nano-titania (NHLT) illustrated in the inset plot.

Figure 3 depicts the results of the FTIR analyses after curing time of 1, 11 and 28 days. In the ML samples without nano-titania, portlandite was identified, exhibiting the characteristic absorption at 3640 cm^{-1}. Furthermore, the absorptions at 990 and 540 cm^{-1}, which are indicative of the neo-formed hydraulic components, are more pronounced in the samples with nano-titania, confirming enhancement of the hydration process.

SEM micrographs of the mortars ML1 without- (Figure 4a) and with nano-titania MLT1 (Figure 4b) after one year of curing corroborate the results of DTA analysis. SEM micrograph of MLT1 (Figure 4b) shows that a dense network of hydraulic amorphous components was formed, providing more elasticity to the mortar matrix. Moreover, characteristic hexagonal crystals of portlandite were located in both SEM micrographs; nevertheless in the case of MLT1 mortar, the portlandite crystals are obviously fewer. The FTIR spectra of the above studied mortars are in full accordance with the SEM micrographs, showing traces of portlandite and enhanced hydraulic compound formation after a one year curing (Maravelaki-Kalaitzaki 2013).

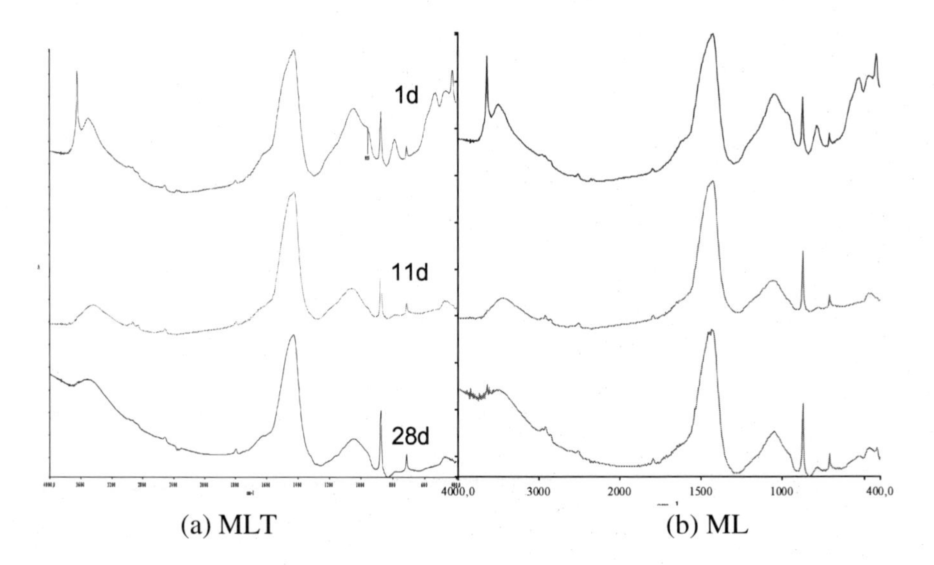

Figure 3. FTIR for metakaolin-lime mortars with (a) and without nano-titania (b) cured for 1, 11 and 28 days. Portlandite (P) still exists in the ML mortars after 28 days of curing.

These variations can be explained by the TiO_2 photocatalytic action for the mixtures with nano-titania; the latter leads both to calcite formation and enhanced hydration of CSH and CAH products, similar to what was observed during the early age hydration of Portland cement by adding increased dosage of fine titanium oxide (Jayapalan 2009).

Figure 4. SEM micrographs of (a) ML mortar and (b) MLT mortar showing portlandite crystals (P) and the hydraulic amorphous components.

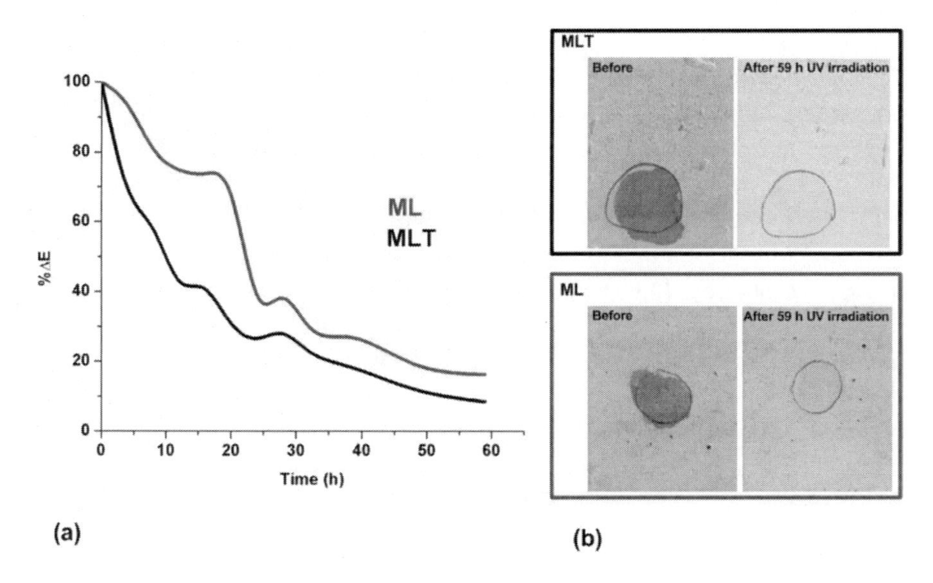

Figure 5. Decrease of ΔE% of stains of methylene blue on mortars with metakalolin-lime and nano-titania (MLT) and metakaolin-lime (ML) after 59 h under UV irradiation.

The photocatalytic properties of the nano-titania mortars were tested by applying both stains of methylene blue and methyl orange to the mortars' surfaces and observing their discoloration. The intensive decrease of the total color difference ΔE% observed on the mortar surface containing nano-titania stained with methylene blue after 59 h of UV irradiation, can be considered as an indication of the potential of nano-titania mortars to decompose organic pollutants (Figure 5).

Mechanical Properties of the Adhesive Mortars and the System Stone-Mortar

Table 3 reports the physical and mechanical properties of the designed mortars. The physical properties of the designed mortars differ insignificantly, indicating that the nano-titania addition neither modified the microstructure, nor affected the hygric behaviour of the materials. The lowest Fc values were recorded for the NHL samples with a B/A=1 (NHLT1), while the Fc values decreased with curing time in the ML samples. Even though, the Fc values recorded at four weeks curing for the MLT samples are lower than the corresponding values for samples without nano-titania (ML), nevertheless, the Fc values of the MLT samples reached higher values than the ML samples after three months and one year curing, thus indicating the beneficial effect of the nano-titania in the compressive strength. The decrease of Fc values over time in the ML compositions has been already reported by other authors (Aggelakopoulou 2011, Velosa 2009) and was most probably attributed to the microcracking formation due to shrinkage during the curing.

By comparing the results of physical and mechanical properties, the mortars NHLT2, ML and MLT were selected as adhesive means for the stones under consideration (Table 4). The mortar NHLT1 showed similar values to the others, except for its high capillary water coefficient; this is probably the reason this mortar exhibited difficulty in joining the stone specimens. As such, it was therefore not included in the finally selected mortars.

It seems that nano-titania with its hydrophylicity created an environment, which not only enhanced the hydraulic component formation, but also controlled the shrinkage, thus avoiding microcracking (Karatasios 2010). Further support to this statement is derived from the dense network of hydraulic components observed in the SEM micrographs (Figure 4b) for the MLT samples.

Table 3. Physical and mechanical properties of the designed mortars

Code	WCC* g cm^{-1} s$^{-1/2}$	WSC* %	P %	Pr μm	SSA m^2/g	Curing	Fc^ MPa	Ff-3pb* MPa	E GPa
NHLT1	0.0149 (±0.007)	27.80 (±1.09)	32.96	0.295	12.0	4w	4.15 (±0.16)	1.16 (±0.05)	0.17
						3m	5.47 (±0.28)	1.69 (±0.04)	0.22
						1y	5.51 (±0.95)	-	0.23
NHLT2	0.0080 (±0.0004)	25.58 (±0.27)	37.79	0.092	12.84	4w	5.41 (±0.29)	1.03 (±0.05)	0.37
						3m	-	-	-
						1y	7.22 (±1.61)	-	0.41
ML	0.0040 (±0.0001)	33.92 (±0.26)	31.46	0.031	14.02	4w	14.85 (±1.53)	1.15 (±0.05)	0.42
						3m	11.66 (±1.26	-	0.76
						1y	10.68 (±1.02)	-	0.56
MLT	0.0074 (±0.003)	29.52 (±1.07)	32.54	0.031	16.01	4w	13.26 (±1.52)	1.21 (±0.05)	0.59
						3m	14.19 (±0.70)	-	1.10
						1y	15.40 (±1.87)	-	0.91

(*) mean value of three samples; (^) mean value of 6 samples; WCC: Water Coefficient by Capillarity; WSC: Water Saturation Coefficient; P: porosity; Pr: average pore radius; SSA: Specific Surface Area; Fc: Compressive Strength; Ff-3pb: 3-point Flexural strength; E: Elasticity Modulus, [4w]: 4 weeks; [3m]: 3 months; [1y]: 1 year; -: n. a.

Table 4 reports the results of the 4-point bending test and the direct tensile test for the adhered stone-mortar specimens (Figure 1). In all tests, failure was observed at the interface between stone and mortar. The results revealed that the higher bonding strength (higher flexural and tensile strengths) was measured when applying the NHLT2 mortar, as compared to the MLT and ML mortar.

Based on the physico-chemical and mechanical characterization of all the studied adhesive mortars the MLT mortars with metakaolin, lime and nano-titania and the natural hydraulic lime mortars with nano-titania (NHLT2) have been selected as the most appropriate for the adhesion of fragment porous stones.

Table 4. Mechanical properties of stone-mortar specimens cured for four weeks

Mortar Code	Number of Ff-4pb adhered stone specimens	Ff-4pb (MPa)	Number of Ft adhered stone specimens	Ft (MPa)
NHLT2	4 (D1, D2)	2.39 (±0.70)	2 (D1, D2)	0.51 (±0.26)
MLT	3 (D3, K)	1.34 (±0.47)	2 (K)	0.15 (±0.08)
ML	1 (D4)	1.07	1 (D1)	0.09

Ff-4pb: 4-point Flexural strength; Ft: direct tensile strength.

CONCLUSION

This research work addressed an important problem in the restoration sector concerning the reassembling of stone fragments from ancient monuments using non-cementitious mortars. The proposed adhesive mortars contain hydraulic lime or metakaolin and lime as binders, carbonate sand with grains between 250 and 63 μm and binder to aggregate ratio 1 or 2. The nano-titania as additive was employed in a binder replacement of 4.5-6% w/w.

The mechanical characterization indicated that the mortars with nano-titania showed increased compressive and flexural strengths and moduli of elasticity over time, when compared to the specimens without nano-titania. The results also indicate enhanced carbonation and hydration of mortar mixtures with nano-titania. The hydrophylicity of nano-titania improves the humidity retention of mortars, thus facilitating the carbonation and hydration processes. This property can be exploited into the fabrication of mortars tailored to adhering porous limestones, where humidity controls the mortar setting and adhesion efficiency. Home-designed equipment applied to measure the bonding strength of stone-mortar systems revealed that the nano-titania addition in both metakaolin-lime and hydraulic lime mortars improves the adhesive property of the mortar when applied to porous stones. The ability of nano-titania mortars to discolor stains of methylene blue under visible light irradiation can be used in favor of the avoidance of biological crusts. Considering all the above and the bonding strength, binders of metakaolin-lime and/or hydraulic lime mortars with nano-titania have been selected to carry out the adhesion of fragments porous stones in restoration interventions.

ACKNOWLEGMENTS

This work was carried out in collaboration with the Akropolis Restoration Service (YSMA). The authors would like to thank Emeritus Prof. Vasileia Kasselouri-Rigopoulou, (Committee for the Restoration of the Acropolis Monuments-ESMA) for collaboration and scientific support, the Committee for the Restoration of the Acropolis Monuments, Dr. E. Aggelakopoulou Head of the Technical Office for the Acropolis Monuments' Surface Conservation and Dr. E. Sioubara, scientific responsible for the reassembling of fragment stones of Archaic period. This research has been co-financed by the European Union (European Social Fund - ESF) and Greek national funds through the Operational Program "Education and Lifelong Learning" of the National Strategic Reference Framework (NSRF) - Research Funding Program: Heracleitus II, Investing in knowledge society through the European Social Fund.

REFERENCES

Aggelakopoulou, E., Bakolas, A., Moropoulou, A. (2011) Properties of lime–metakolin mortars for the restoration of historic masonries, *Applied Clay Science,* 53 (1), 15–19.

Amstock, J.S. (2000), *Handbook of adhesives and sealants in construction,* McGraw-Hill, New York.

Hyeon-Cheol, J., Young-Jun, L., Myung-Joo, K., Ho-Beom, K., Jun-Hyun, H. (2010) Characterization on titanium surfaces and its effect on photocatalytic bactericidal activity, *Applied Surface Science,* 257 (3), 741–746.

Jayapalan, A.R., Lee, B.Y., Kurtis, K.E. (2009), *Effect of Nano-sized Titanium Dioxide on Early Age Hydration of Portland Cement, Proceedings of the NICOM3, Nanotechnology in Construction.*

Karatasios, I., Katsiotis, M.S., Likodimos, V., Kontos, A.I., Papavassiliou, G., Falaras, P., Kilikoglou, V. (2010) Photo-induced carbonation of lime-TiO_2 mortars, *Applied Catalysis B: Environmental,* 95 (1-2), 78-86.

Maravelaki-Kalaitzaki, P., Agioutantis, Z., Lionakis, E., Stavroulaki, M., & Perdikatsis, V. (2013) Physico-chemical and Mechanical Characterization of Hydraulic Mortars Containing Nano-Titania for Restoration Applications, *Cement and Concrete Composites,* 36 (1), 33-41.

Moropoulou A, Bakolas A, Moundoulas P, Aggelakopoulou E, Anagnostopoulou S. Strength development and lime reaction in mortars for repairing historic masonries. *Cem Concr Compos* 2005; 27:289-94.

Papayianni, I., Stefanidou, M. (2006) Strength–porosity relationships in lime–pozzolan mortars, *Construction and Building Materials,* 20 (9), 700-705.

Rosario Veiga, M., Velosa, A., Magalhaes, A. (2009) Experimental applications of mortars with pozzolanic additions: Characterization and performance evaluation, *Construction and Building Materials,* 23 (1), 318-327.

Tziotziou, M., Karakosta, E., Karatasios, I., Diamantopoulos, G., Sapalidis, A., Fardis, M., Maravelaki-Kalaitzaki, P., Papavassiliou, G., Kilikoglou, V. (2011) Application of ^1H NMR in hydration and porosity studies of lime pozzolan mixtures, *Microporous and Mesoporous Materials,* 139 (1-3), 16-24.

Velosa, A.L., Rocha, F., Veiga, R. (2009) Influence of chemical and mineralogical composition of metakaolin on mortars characteristics, *Acta Geodynamica et Geomaterialia,* 153 (6), 121-126.

Whittaker, B., Singh, N., Sun, G. (1992) *Rock Fracture Mechanics: Principles, Design and Applications,* Elsevier, ISBN 9780444896841, pp. 145-160.

In: Adhesives
Editor: Dario Croccolo
ISBN: 978-1-63117-653-1

Chapter 5

FATIGUE CRACK GROWTH MONITORING OF BONDED COMPOSITE JOINTS

Andrea Bernasconi*[*] *and Md Kharshiduzzaman
Politecnico di Milano, Dipartimento di Meccanica, Milano, Italy

ABSTRACT

In modern engineering applications, use of bonded composite joints is becoming increasingly widespread due to the numerous advantages over conventional mechanical fastening. For load bearing applications, fatigue cracks, either induced by defects or by applied stresses, may appear and propagate, thus becoming potentially harmful for the structural integrity of a part or a whole structure. Therefore, in-situ structural health monitoring (SHM) of bonded joints is essential to maintain reliable and safe operaional life of these structures. This chapter presents various in-situ SHM techniques which are used to monitor bonded composite joints with a focus on fatigue crack monitoring based on the backface strain (BFS) technique. Case studies are presented and discussed for adhesively bonded single lap joint (SLJ). Sensors associated with this technique are also explained, with emphasis on Fiber Bragg Grating (FBG) optical sensors.

[*] Andrea Bernasconi, Associate Professor, Department of Mechanical Engineering, Politecnico di Milano, Via La Masa 1, 20156 Milano, Italy Email: andrea.bernasconi@polimi.it; Tel.: +39-02-2399-8222 ; fax: +39-02-2399-8263.

Keywords: Structural Health Monitoring (SHM), Composite material, Bonded joint, Back Face Strain (BFS) technique, Fiber Bragg Grating (FBG) sensor, Single Lap Joint (SLJ)

INTRODUCTION

Composites have emerged as important materials in the last few decades because of their high specific stiffness and high specific strength, allowing for lighweight design solutions; moreover, they possess excellent fatigue resistance and outstanding corrosion resistance compared to most common metallic alloys, such as steel and aluminum alloys. Besides, composites include the ability to fabricate directional mechanical properties, low thermal expansion properties and high dimensional stability. It is the combination of outstanding physical, thermal and mechanical properties that makes composites attractive to use in place of metals in many applications, particularly when weight-saving is required [1, 2].

Bonded joints are extensively employed in the construction of composite structures in aerospace applications, maritime structures, lifting equipment, wind mills as well as automotive industries [3, 4, 5]. Unlike the bolt hole in mechanical fastening that causes a stress concentration in the composite joint plates, adhesively bonded joints minimize the potential for stress concentration within the joint. Besides, applications where lower structural weight, improved damage tolerance design philosophy are required, adhesively bonded joints provides a potential solution. Bonded joints are an efficent fastening solution also for hybrid structures, i.e., structures where composite parts are connected to metal parts.

In the case of high strength joints, difficulties arise because manufacturing defects and operational loads (static, cyclic fatigue loads or impacts) can cause local debonding. Therefore, to maintain damage tolerance design philosophy in cases of composite structures as well as increasing safety and decreasing cost service and maintenance purposes, developing structural health monitoring (SHM) technique for evaluating and monitoring critical bonded composite joints is extremely important.

In this chapter, various SHM techniques applicable to monitoring the structural integrity are presented and discussed. Special attention is given to the SHM technique to monitor fatigue crack propagation based on strain field measurement. Basic principle of strain based SHM method is explained and some case studies, referring to this technique applied to bonded joints using both electrical resistance strain gauge and Fiber Bragg Grating (FBG) optical sensor, are predented.

ADHESIVELY BONDED COMPOSITE JOINTS

Joint Definition and Configuration

In adhesive bonding process, material, termed as substrate, is joined by placing and solidifying an adhesive between the adherend surfaces to produce an adhesive joint. It is also possible to join similar as well as dissimilar materials by adhesively bonded joint. Proper surface preparation must be taken care of the substrates prior to bonding for better durability of the joint [6].

In case of bonded composite joints, the non-uniform stress distribution along the bonding surface should always be accounted for. The peak stress is mainly dependent on the bonding pattern of the joint, adhesive thickness, bonded length, joint geometry, adherend stiffness imbalance, ductile adhesive response, and the composite adherends.

A wide variety of adhesively bonded joint configurations are available to the designer based on the application and scope of use [7]. Commonly, joint configurations that have been analyzed in the literature are single-lap joints, double-lap joints, scarf joints, and step joints (Figure 1).

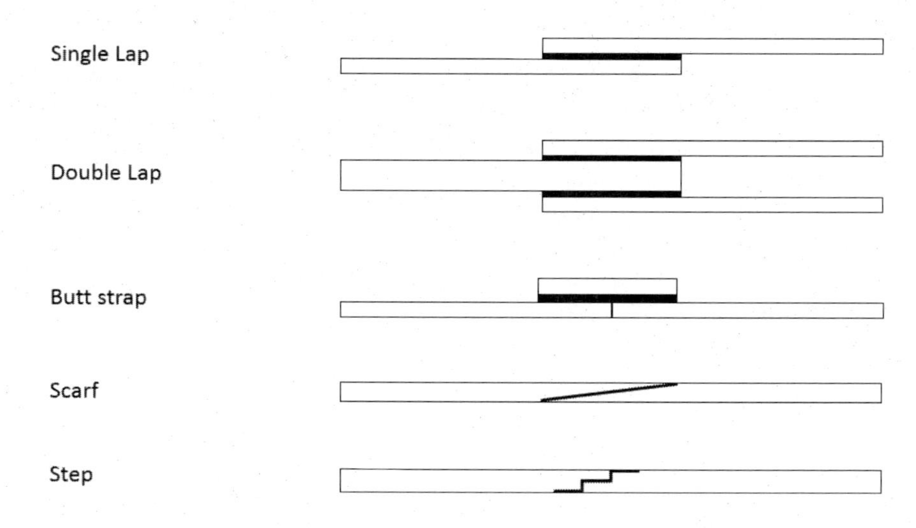

Figure 1. Various joint configurations for adhesively bonded joints.

The strength of a given type of joint depends, for a given type of load, on the stress distribution within the joint, which in turn depends on the joint geometry and the mechanical properties of adhesive and adherend. Moreover, the strain field in the vicinity of the joint depends on the joint configurations, thus making it possible, in some cases, to find a correlation between modifications of the strain field and the integrity of the adhesive joint.

Problems Associated with Bonded Composite Joint

Composite materials as well as adhesively bonded joints are not exempt from some problems even though they have great advantages and applicability. Defects related to composite bonded joints can be classified into two main streams- manufacturing defects and in service defects.

The source of manufacturing defects in a bonded composite joint can be due to defects during manufacturing the composites and during manufacturing the joint while bonding. The main composite manufacturing defects are delamination, resin starved areas, resin rich areas, blisters, air bubbles, voids, wrinkles, thermal decomposition [8]. On the other hand, while manufacturing adhesively bonded joints, most defects are to be found either at the interface between the adhesive and the adherent or within the adhesive (Figure 2) which can lead to poor durability of the joint.

Figure 2. Typical crack propagation in a composite bonded joint.

Surface preparation is also very important for long term durability of bonded joints. Lack of proper surface preparation of substrates prior to adhesive bonding can attribute to surface contamination which leads to adhesive failure.

Apart from the manufacturing flaws, there are a number of in-service causes that can cause failure to bonded composite structures, such as environmental degradation, impact damage, fatigue, and cracks from local overloading from high static loads. Moreover, improper joint selection and design in bonded joints can result in a number of failure modes, including shear or peel failure in the adhesive; shear or peel in composite surface plies. The failure also depends on the joint configuration [9].

Furthermore, the mechanisms of load transfer from one adherend to another through adhesive layer, flaw initiating, spreading over the composite volume and leading to the ultimate failure can be very complicated, as shown in Figure 2, where a typical crack propagation in a composite bonded joint is shown, ultimately ending into delamination of the composite substrates. Therefore, clear description for the damage evolution and fracture behavior in composites remains a challenge work at present days. Hence, the structural health monitoring of adhesively bonded composite joints remains critically important for maintaining safe, damage tolerant design philosophy.

Structural Health Monitoring (SHM) of Bonded Composite Joints

Structural Health Monitoring (SHM) can be defined as the method by which sensors are utilized to identify and quantify structural integrity through a continuous in-service monitoring process. In this manner, corrective maintenance can be performed when required before the structure reaches the critical point of catastrophic sudden failure.

Even though adhesively bonded composite joints have a number of advantages over conventional mechanical fastening, they have some drawbacks too. Adhesively bonded joint is a permanent joint and it can't be disassembled for maintenance and damage inspection. Therefore, it is highly essential to monitor the state of the joint throughout its service life in order to prevent drastic failures, to optimize use of the structure which also identifies the scope of future product development and improvements.

There are a number of techniques in the field of SHM for analyzing structural integrity of bonded composite joints. Most of the conventional monitoring systems, such as microscopic failure analysis, ultrasonic, X-ray, thermography and eddy current method are off-line techniques [10, 11], i.e., it's not possible to monitor real time during the service life. But for adhesively bonded joints, real time in-situ monitoring is required. SHM techniques which enable the possibility of in-situ monitoring include acoustic emission, carbon nanotubes network, active vibration method and backface strain based technique.

Acoustic Emission (AE)

Acoustic emission consists of a rapid release of energy as a material undergoes fracture or deformation. In SHM, two important features of acoustic emission are important: its ability to perform remote detection of crack formation or movement at the time it occurs and to do this continuously [12].

A schematic representation of acoustic emission (AE) technique is shown in Figure 3. When a crack initiates in a component under applied load, the crack acts as a source of AE waves which propagates in all direction. Sensor attached on the surface of the component records the AE waves and sends signals to the acquisition system for diagnosis.

It is important to note that, AE waves generated at the damage location will have to propagate over a certain distance before detection takes place at the sensor's location, and therefore, the shape of an AE wave can be altered due to attenuation, reflection, mode conversion, and dispersion. It is highly recommended to have a good understanding of these effects as they can have a significant effect on the AE signals obtained during a test.

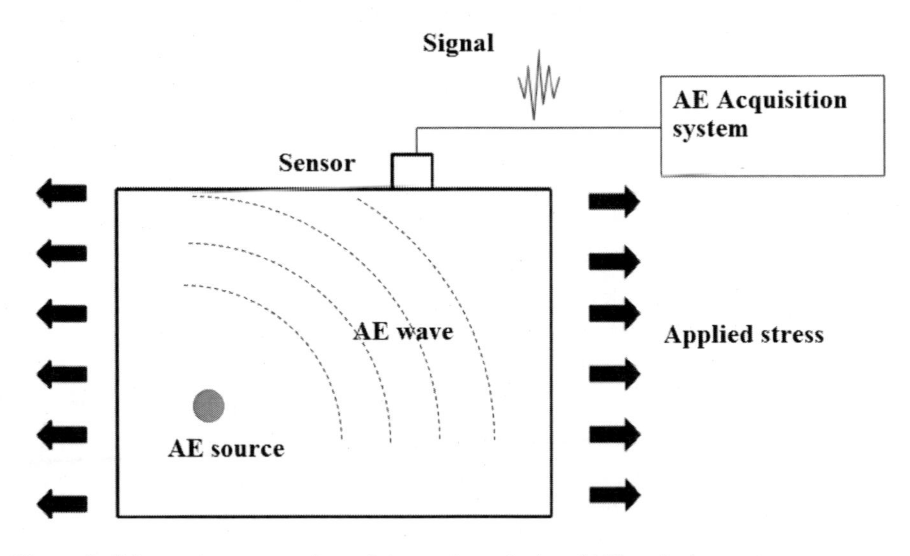

Figure 3. Schematic reprentation of Acoustic emission (AE) technique.

As the mechanical properties of a composite structure are strongly dependent on the direction along which they are measured, this in turn will result in wave propagation effects that vary from one direction to another. For example, in a unidirectional laminate, waves will generally propagate faster along the fiber direction than along the direction perpendicular to the fibers. These effects have to be taken into consideration.

Ducept et al., [13] studied mixed mode failure criteria for a glass/epoxy composite and an adhesively bonded composite joint. In their study, the initiation failure point detected by acoustic signal and by the non-linearity point on the load/displacement curve and found good correspondence. Magalhaes et al., [14] studied the application of acoustic emission to investigate the creep behavior of composite bonded lap shear joints.

Carbon Nanotubes (CNTs) Sensing Networks

Amanda et al., [15] established a technique which holds promise for detecting damage and evaluating different failure modes in adhesively-bonded composite-to-metal single-lap shear joints. They compared electrical resistance data using carbon nanotubes in adhesive with AE and found good agreement.

Carbon nanotubes (CNTs) have found many applications due to their high aspect ratio, high specific stiffness and strength, and excellent electrical and thermal conductivities. In addition to these, carbon nanotubes have enabled their use in sensors and actuators [16, 17].

As a result of their high aspect ratio and electrical conductivity it has been established that carbon nanotubes can form electrically conductive networks in epoxy adhesives and polymer matrix materials and ultimately make them electrically conductive which triggers the opportunity to develop in-situ SHM method. Another important factor is that the addition of carbon nanotubes served to increase the bonding strength and durability of epoxy joints [18].

Thostenson et al., [17] demonstrated that the electrical conductive network formed by CNTs in an epoxy enables to detect the onset, nature and evaluation of damage. In their tests on epoxy/nanotube specimen, they observed that, the electrical resistance of the specimen has linear relationship with the specimen deformation and sharp increase of resistance due to crack initiation.

In another study, Thostenson et al., [19] studied the effect of nanotube concentration on the electrical properties of epoxy and established that percolating networks can be formed at concentrations below 0.1 wt.%. In another investigation, the effectiveness of a carbon nanotube network in quantitatively measuring the onset and progression of damage in real time in-situ monitoring of a reinforced composite is demonstrated [20]. Nofar et al., [21] reported the sensing capability of CNT network in detecting the failure region in laminated composite subjected to static and dynamic loading. They concluded that the CNT network is more sensitive in detecting and predicting the cracked regions than strain gauges due to existence of CNT network throughout the structure as whole rather than locally attached strain gauges. In a more recent study, Amanda et al., [15] suggested an SHM technique by CNT networks to sense and detect damage and evaluating different failure modes in adhesively-bonded composite-to-metal single-lap shear joints.

Active Vibration Based Technique

Active vibration-based monitoring method is a classical SHM technique. The main idea behind the method is that structural dynamic characteristics are functions of the physical properties, such as mass, stiffness and damping [22]. Hence, physical property changes due to damages can cause detectable differences in vibration responses. The dynamic characteristic parameters usually used in the technique include frequency, mode shape, power spectrum, mode curvature, frequency response function (FRF), mode flexibility matrix, energy transfer rate (ETR), etc.

Chiu et al., [23] demonstrated a monitoring system, which provides information about the in-service performance of the repair and the associated structure. The idea was based on the detection of debonding in the adhesive layer between the repair and the metallic parts. They also proposed that piezoceramic sensors could be used in the system for monitoring because of easy application, economical and reliable. Mickens et al., [24] developed a vibration-based method of damage detection system for monitoring ageing aircraft structures. The method was based on the detection of damage during operation of the aircraft before the damage propagation and the catastrophic failure. In this technique, four piezoelectric patches were used alternatively as actuators and sensors in order to send and receive vibration diagnostic signals. White et al., [25] performed experimental investigations on representative adhesively bonded composite scarf structures. Piezoelectric devices were used to excite and measure the response of the repaired structure. The frequency response of the repaired structure with simulated debondings was found to differ significantly from that one with undamaged repaired one, which signified the application of vibration based of a structural health monitoring system for bonded composite repairs and joints. In another work, White et al., [26] described the development of an SHM technique based on frequency responses for the detection of debonding in composite bonded patches. Two commonly applied repair schemes, the external doubler repair and the scarf repair, were investigated. It was verified that damage could be detected through changes in the frequency responses of the repairs with and without defects for both types of bonded repair schemes.

Strain Based Techniques

Strain-based methods are effective SHM methods, because the presence of damage in the structure under operational loads can alter the local strain distribution due to the load path changing within the structure. Both electrical resistance strain gages as well as Fiber Bragg grating (FBG) optical sensors are usually used in the method to measure the strains. Due to the point nature of these sensors, strain monitoring is more suited to local structural 'hotspots' in large components rather than wide-area monitoring when individual or small no. of sensors are to be used. Nevertheless, thanks to the use of an array of sensors [27, 28, 29], as well as the multiplexing capability of FBG sensors, much wider area can be monitored, too.

The strain-based method can be performed by measuring the strain distribution of the intact structure in advance, which acts as the reference. Damage can be then detected when the strain measurement of the structure under loading condition significantly diverges from the reference strain.

FBG sensors provide a potential solution for strain based SHM techniques especially for composites and bonded structures. In composite materials, FBGs can be embedded inside the composite texture with a negligible effect on the mechanical properties [30, 31, 32, 33]. FBG strain sensors have been used in order to detect microscopic damages in composite laminates [34] and also for the investigation of delamination detection in CFRP laminates [35].

FBG sensors have been used to monitor crack growth in composite joints [36, 37]. Bernasconi et al., [27] investigated fatigue crack growth in adhesively bonded joints of thick composite laminates experimentally using an array of equally spaced FBG sensors. Sensors were applied to the side of a single lap tapered joint, in order to simulate its embedment into the composite laminate. They showed that this configuration of FBG array can detect and monitor a fatigue crack propagating in the adhesive joint.

Schizas et al., [38] used an array of five FBG sensors embedded between two unidirectional composite laminate plates near the adhesive bonding in a composite-Makrolon SLJ in order to investigate and monitor non-homogeneous strains. They concluded that FBGs can be used efficiently in non-homogeneous strain environments present in joints of dissimilar materials and delaminations and very useful in testing and monitoring durability.

One very efficient strain based technique for SHM is Backface strain (BFS) measurement in order to monitor crack initiation and propagation under both static and fatigue loading in adhesive joints. In this method, sensors are placed onto the exposed surface i.e., backface of the material to monitor the strain. This technique and sensors associated with this method is described in detail in the following section.

BACKFACE STRAIN (BFS) TECHNIQUE FOR SHM

Backface strain (BFS) measurement is a technique that can be used for the in-situ monitoring for SHM of crack initiation and propagation in adhesively bonded composite joints. Zhang et al., [39] assessed the use of the BFS technique in bonded joints. Their work with single lap joint (SLJ) allowed for finding out that by detecting the switch in the direction of the backface strain crack initiation can be monitored.

The work of Crocombe et al., [40] regarding BFS technique concluded that the BFS response was highly dependent on the location of the sensor and the greatest change in backface strain with damage occurs in the overlap zone. Therefore, the position of the strain sensor should be inside the overlap because here it will produce the largest change in BFS with damage. Shenoy et al., [41] investigated the crack initiation and propagation behavior in SLJ using BFS technique and showed that fatigue failure in adhesively bonded joints goes through various stages, starting from an initiation period, then a slow crack growth period followed by a faster growth period before a rapid quasi-static fracture.

Solana et al., [42] investigated the fatigue life of adhesively bonded joints using backface strain technique. They used six electrical strain gauges (SGs) placed along the overlap, 3 on each end of the substrates to monitor fatigue initiation and propagation within the adhesive layer. This configuration was able of discriminate the damage profiles at both ends, showing where it initiated and which side was damaged as a consequence.

As BFS technique is based on sensing the pattern of strain of the backface, strain sensors are needed for the sensing. Both conventional electrical resistance strain gages as well as FBG sensors can be used for BFS techniques.

Electrical Strain Gauges

An electrical strain gauge mainly consists of a resistance grid of very fine wire or more commonly, metallic foil in a grid pattern, connectors (also called solder tabs) and an encapsulation layer. The grid pattern maximizes the amount of metallic wire or foil subject to strain in the parallel direction. The cross-sectional area of the grid is minimized to reduce the effect of shear strain and Poisson strain. When the object undergoes deformation, the foil is deformed, causing its electrical resistance to change. This change of resistance, usually measured using a Wheatstone bridge, is related to the strain by the quantity known as the gauge factor. Gauge factor is defined as the ratio of fractional change in electrical resistance to the fractional change in length (strain). To connect the strain gauge to the Wheatstone bridge, a minimum of two electric wires is required, with an optional third wire for minimization of the effect of wire resistance upon the strain measurements.

Fiber Bragg Grating (FBG) Optical Sensors

Fiber Bragg Gratings (FBGs) are becoming increasingly more popular for strain based SHM technique. A fiber Bragg grating is region of periodic refractive index modulation inscribed in the core of an optical fiber such that it diffracts the propagating optical signal at a specific wavelength. Bragg grating is written into a single mode optical fiber. These fibers consist of a very small diameter inner core, between 4 to 9 μm, and outer part, called cladding, having diameter of 125 μm. Due to doping by Germanium, the core has higher refraction index than the outer part, and therefore, light propagates only through the inner core [43].

A strong ultra-violet (UV) writing laser light, which is split into two equal intensity beams, later recombined on to the fiber after traversing through two different optical paths. This forms an interference pattern at the core of a photosensitive fiber and thus creating the grating in the core of the optical fiber, which acts essentially as a wavelength-selective mirror [44]. The grating reflects the light at the Bragg wavelength (λ_B), which can be defined as-

$$\lambda_B = 2\Lambda n_{eff} \tag{1}$$

where Λ is the grating period and n_{eff} is the effective index of the fiber core. Any change of the grating period due to strain results in changing the characteristic wavelength of the reflected spectrum. Strain is measured based on this shift of the wavelength, Δλ, of the reflected spectrum, which can be calculated as-

$$\varepsilon_m = \frac{1}{k} * \frac{\Delta\lambda}{\lambda_0} - \left(\alpha_{sp} + \frac{\alpha_\delta}{k}\right) * \Delta T \ \ldots \quad (2)$$

where, ε_m= mechanically caused strain, $\Delta\lambda$= wavelength shift, λ_0= wavelength of the reflected spectrum at unstrained condition, k= gauge factor (0.78), α_{sp}= expansion co-efficient per K of the specimen, α_δ = change of refractive index and ΔT = temperature change in K.

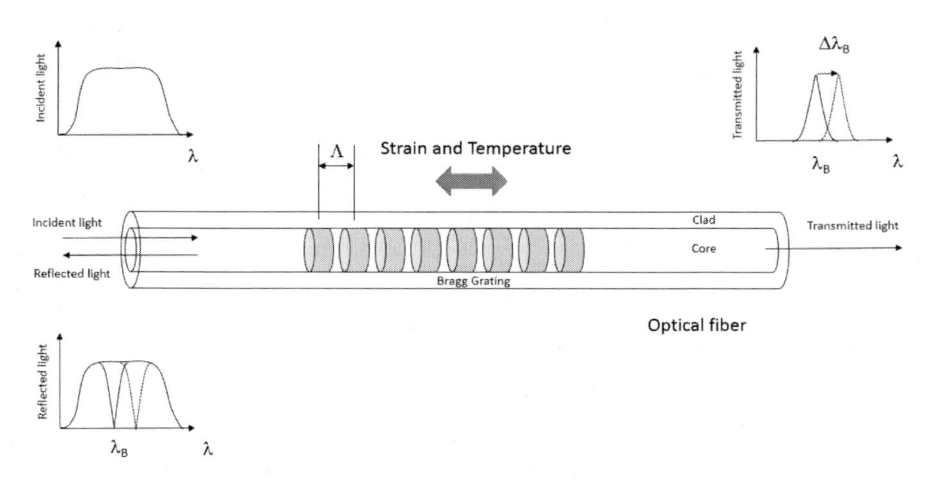

Figure 4. Fiber Bragg grating sensor working principle.

FBG sensors have many advantages, such as higher adaptability to composite materials, being it possible to insert the fibers into laminates with little effects upon their strength and the stiffness. Moreover, they are suitable to measure very high value of strain, are independent from interference of the electromagnetic type, and have good resistance to corrosion, high long term stability, small size and a light weight. The signal produced by FBG sensor is independent of the distance. Besides, FBG sensors are intrinsically passive, so no electrical power is needed and therefore, can be applied to high voltage and potentially explosive environments.

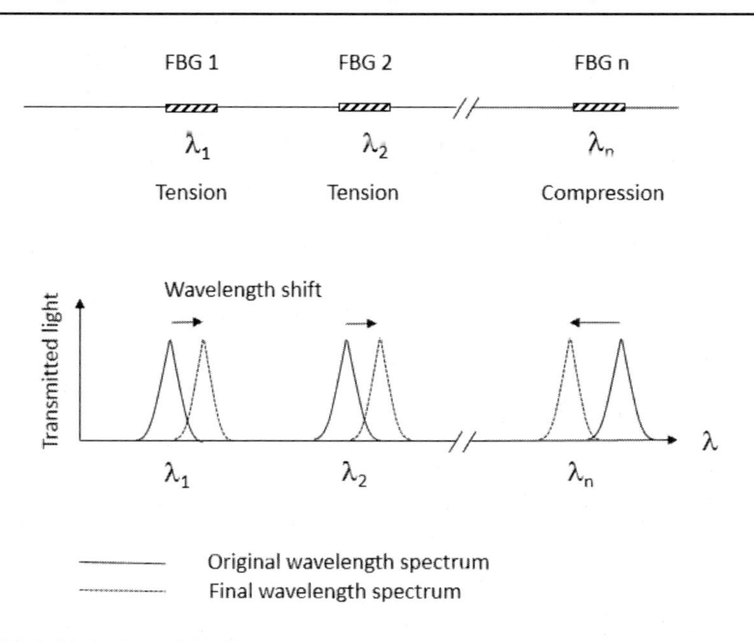

Figure 5. Multiplexing of FBG sensors.

Another unique capability of FBG sensor is the multiplexing [45], i.e., a number of FBG sensors can be inscribed into a single optical Fiber. Each individual FBG sensor reflects back a spectrum centered in a wavelength different from one another (Figure 5), thus creating an array of strain sensors suitable to monitor large structures.

In parallel to the great advantages, FBG sensors have some features which must be taken into account while considering their application. FBG sensors show great temperature dependence which is evident in equation 2. The shift of the wavelength for 1°C is almost equal to the shift produced by an 8 μm/m strain [43]. Besides, FBG shows high sensitivity to lateral force and pressure and the stiffness of FBG sensors causes higher parallel force to the specimen. Nevertheless, in spite of these aspects, FBG is showing great potential for strain sensing due to its unique advantages.

Case Studies on Backface Strain Technique for Fatigue Crack Growth Monitoring

Adhesively bonded SLJs have been extensively studied by numerous researchers due to their simplicity and efficiency [47]. One feature associated to this joint configuration is the stress distribution (shear and peel) in the bondline, which is characterized by stress concentrations at the ends of the bondline. Under these conditions, fatigue crack nucleation at one or both ends of the overlap is very likely. Among the different SHM techniques for monitoring structural integrity of adhesively bonded joints, we investigated the BFS technique. In BFS technique, monitoring of structural integrity can be performed using arrays of sensors, thus allowing for the understanding of the BFS distribution in the presence of a crack and during crack propagation.

In the following sections, two case studies of SHM based on BFS technique for SLJ are summarized (details can be found in [28, 46]) and discussed in terms of:

- the finite element analyses to understand BFS behaviour in SLJ,
- the relationship between BFS distribution and crack position during crack propagation,
- the formulation of crack monitoring correlation, and
- the use of an array of strain sensors to capture BFS distribution for experimental verification of SHM technique.

Case 1 refers to a tapered carbon fiber SLJ and case 2 refers to a hybrid aluminum-carbon fiber reinforced polymer SLJ.

CASE 1: TAPERED SINGLE LAP JOINT

Specimen

In the first case [28], a tapered lap joint was considered having two CFRP composite laminate thick bars as adherents. The shape and dimension of the specimen is given in Figure 1(a). Total thickness of the laminate is 10 mm. The laminate is made of woven (W) and unidirectional (UD) laminae of CF and epoxy matrix having the stacking sequence of $[45_W / 0_{5\ UD} / (0/45_2/ 0_2/ 45_2/ 0)_W / 0_{5\ UD} /45_W]$. The adopted adhesive was the 3M Scotch-Weld 9323 B/A bi-component epoxy.

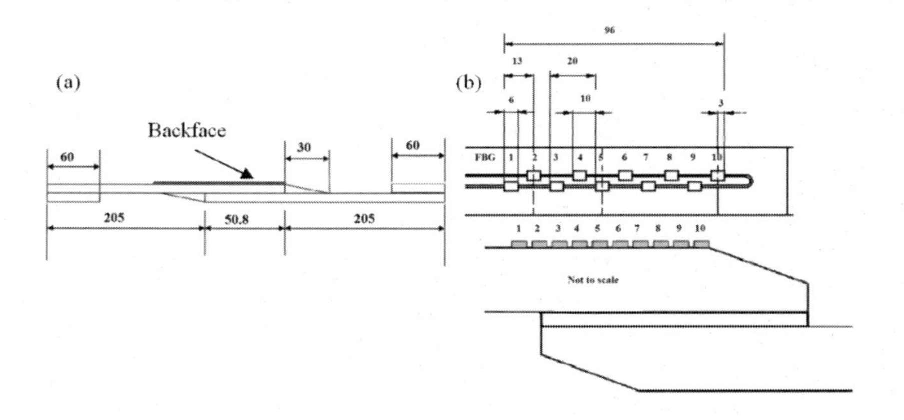

Figure 6. Detailed dimensions of the tapered SLJ (in mm).

Finite Element (FE) Analysis

The full laminate was modeled in FE analysis using 2D solid elements with partitions having the properties of the corresponding plies. A thin layer of solid elements was inserted between the adherends, to simulate the presence of the adhesive layer. By gradually removing elements in the adhesive layer, the crack propagation was modeled. The variation of the BFS (indicated by red line in Figure 6) was extracted for each crack length. The BFS profile has a minimum negative peak and the corresponding position of that negative peak coincides with the crack position (Figure 7). In order to take into account the effect of the non-uniform strain field upon the response of the FBG sensors, the T-matix technique [48, 49] was also applied. Other causes of non-uniformity of the strain field can be the macrostructure of the composite material, particularly in the case of woven laminates [50]. In this case, size and position of sensors should be chosen according to the characteristics of the material's macro-structure, in order to minimize these effects.

Use of an Array of FBG Sensors

In order to capture the BFS distribution along the backface of the bonded region (as shown by red line in Figure 6), an array of FBG sensors is well suited to this purpose. An array of 10 FBG sensors, as shown in Figure 6, was used in this case in order to sense BFS distribution.

Experimental Test

A Schenck servo-hydraulic testing machine with a capacity of 250 kN was used to apply tensile pulsating load to the specimen. This fatigue test caused the crack nucleation and propagation parallel to the bondline. After every 1000 cycles, the test was stopped, the static measurements were recorded with the FBG sensors by swept laser interrogator, MicronOptics sm130-700 and an optical microscope was used to observe the crack position, analyzing images like that of Figure 2.

Results

A comparison between FE simulations and FBG measurements was made. The result is shown in Figure 7 for a crack length of 7 mm. The BFS distribution measured by an array of FBG sensors and interpolated by spline interpolation has a similar trend as the FEM results, as shown in Figure 7.

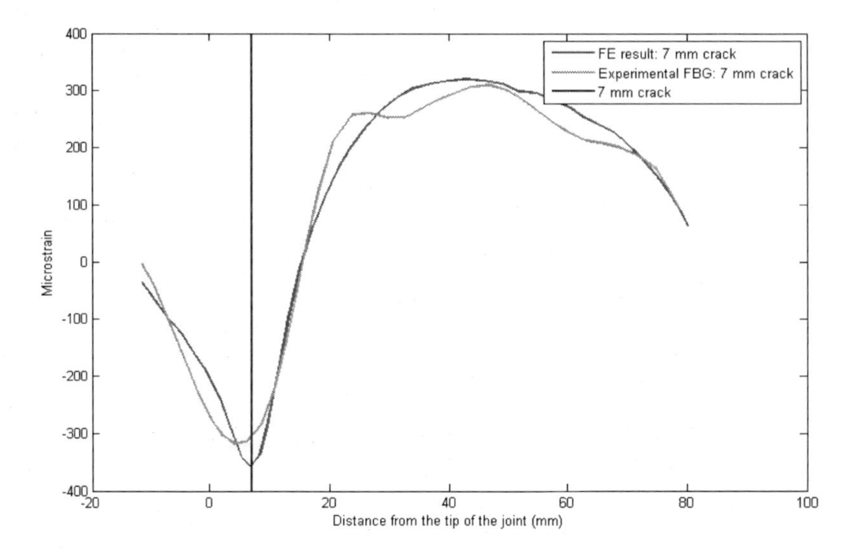

Figure 7. BFS distribution Comparison between FE analysis and FBG measurements for 7.0 mm crack length.

Although the position of the negative peak of BFS profile inferred from FBG measurements does not exactly match the crack position, presumably due to the presence of non-homogeneous strain field in woven composite, this approach showed great potential for crack detection. In facts, during crack propagation, a linear correlation was found between the position of the crack tip measured by optical microscope observation and the position of the negative peak of the BFS distribution.

CASE 2: METAL-COMPOSITE SINGLE LAP JOINT

Specimen

In this case, an aluminum-carbon fiber reinforced polymer (CFRP) adhesively bonded SLJ was investigated [46]. Woven CFRP adherent having 0° ply orientation was bonded with an aluminum adherend by epoxy structural adhesive DP 760.

The detailed shape and dimensions of the specimen are given in Figure 8. Four specimens were manufactured. Two specimens were without any artificial defect and each of the other two specimens had an artificial crack which was created in one side of the joint by inserting a Teflon tape into the adhesive.

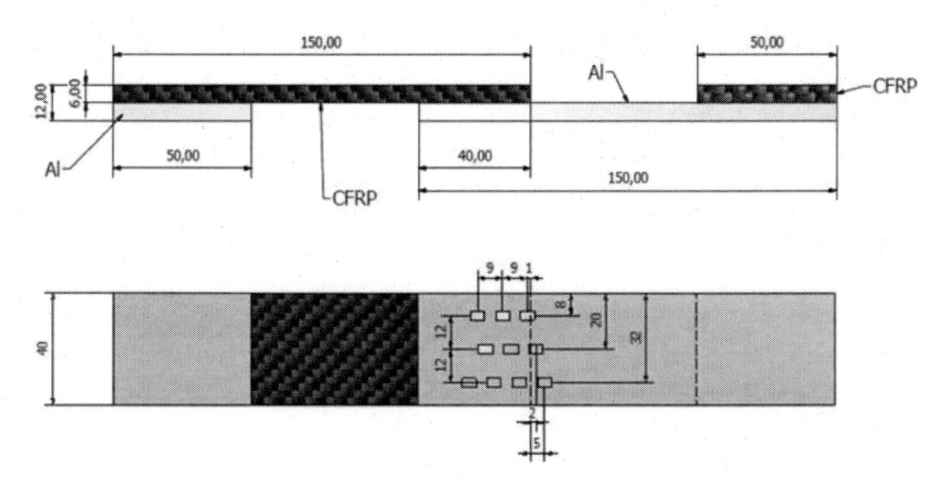

Figure 8. Detailed dimensions (in mm) of the specimen.

Application of an Array of Strain Sensors

In this case, two types of sensors were applied in order to reconstruct the strain distribution along the backface of aluminum substrate of the joint. Both electrical strain gauges (SG) and FBG sensors were used on different specimens. Strain gauge arrays were used for two specimens, one without any defect and another with an artificial of 5.7 mm. FBG arrays were used for other two specimens, one without any crack and the other with a pre-crack of 10 mm.

Due to dimensional restrictions and to leave space for wiring of strain gauges, 3 strain gauge arrays containing total 10 strain gauges were used in the bonded region as shown in Figure 8. In the case of FBGs, due to their multiplexing capability, an array of 10 FBG sensors inscribed along a single fiber was used. While installing the FBG array, the positional distance between the sensors was kept equal to that between the electric strain gauges.

FE Analysis

The BFS distribution was then investigated in FE analyses for the backface of aluminum substrate. A number of FE analyses were carried out by continuously varying the crack lengths in order to understand the behavior of BFS distribution during fatigue crack propagation. The strain values extracted from the FE analyses to construct the BFS profile is shown along the black bold lines (marked in orange arrow) shown in Figure 9(a) in order to match the installed strain sensor configuration as shown in Figure 8. From the analyses for different crack lengths, it is observed that the position of the minimum peak of the backface strain values follows the crack position with a 1.98 mm offset (Figure 9(b)), i.e., also in this case a linear relationship exists between the position of the crack tip and the location of the minimum peak of BFS distribution.

Experimental Investigation

A servo-hydraulic system from MTS Landmark™ Testing Solutions with a capacity of 100 kN was used, in an environment with controlled temperature and humidity. A static load was applied to the specimens with strain gauges and specimen with FBG array without any crack.

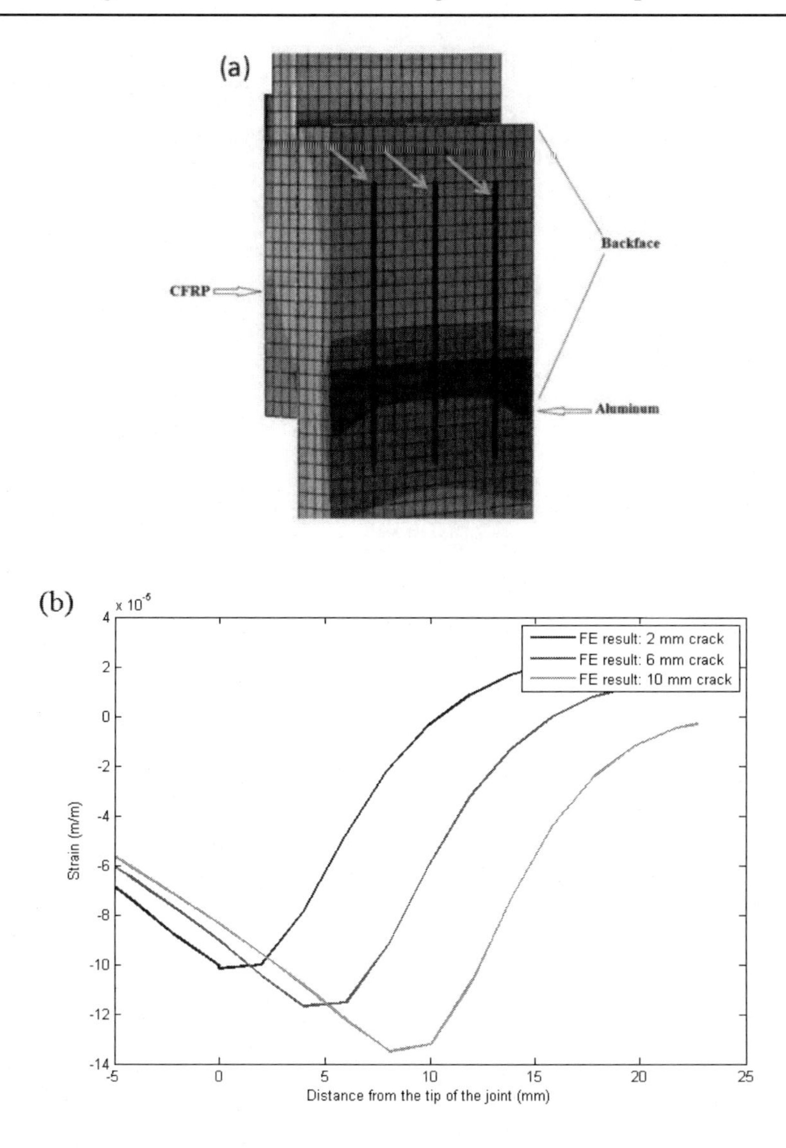

Figure 9. (a) FE model with three lines to be analyzed for BFS distribution, (b) BFS distribution from FE analyses for different crack lengths.

Similarly, a static load was applied to the specimen with strain gauges with a crack of 5.7 mm and to the specimen with FBG array having a pre-crack of 10 mm.

The conditioning of FBG sensors was done using an HBM DI410 interrogator (4-channel device, 1000 measurements/sec). Static investigations were first done in order to check the crack detection technique based on minimum negative peak of the BFS distribution. Then a fatigue test with a load ratio equal to 0.1 was carried out with one of the pre-cracked specimens.

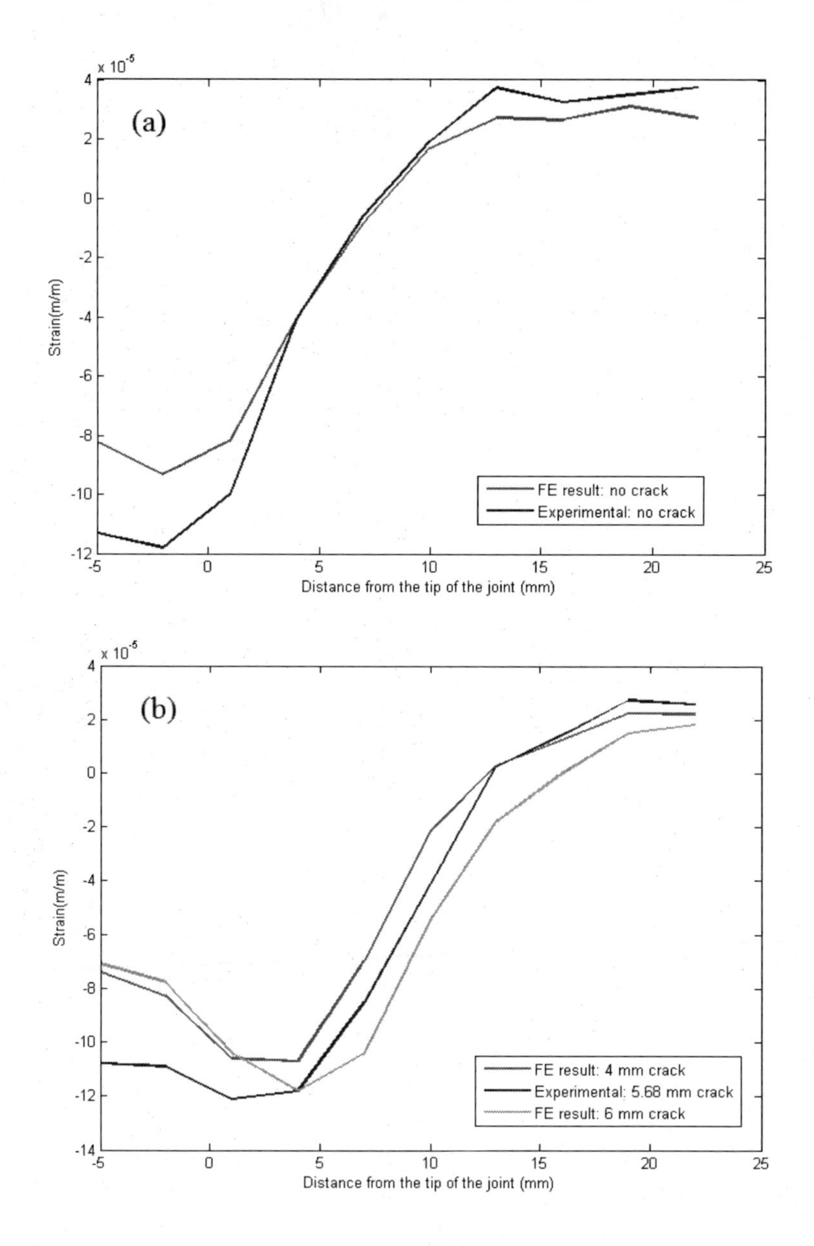

Figure 10. (Continued).

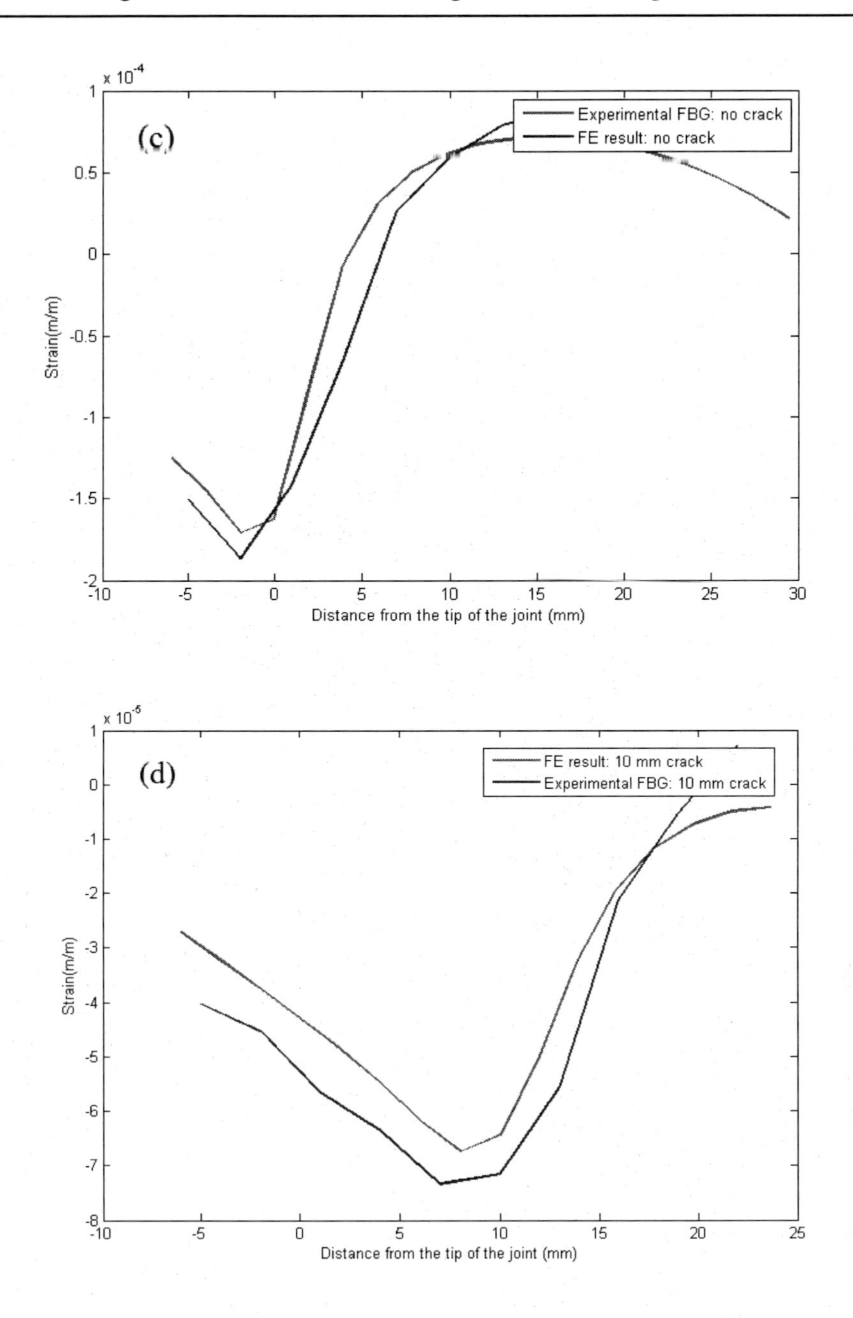

Figure 10. Static results of BFS distribution- by arrays of strain gauges for (a) no crack specimen, (b) specimen with 5.68 mm pre-crack; by an array of FBG sensors for (c) no crack specimen, (d) specimen with 10 mm pre-crack.

The artificial crack was allowed to propagate for 150,000 cycles. During the test, a Matlab function, based on the correlation between crack tip position and the minimum of the BFS distribution obtained by FE analyses, evaluated the position of the crack tip. The propagation of the crack was also monitored by means of an optical microscope.

Result

From the static investigations, as shown in Figure 10, it can be seen that the position of the negative peak of BFS distribution, determined by the experiments using arrays of strain gauges as well as FBGs, matched with the value obtained from FE analysis. In case of the use of electric strain gauge arrays, the magnitude of the BFS value at the negative peak seems to be a bit higher than the FE calculated values, while the FBG array showed greater accuracy.

While carrying out the fatigue test, after 150,000 cycles, a 19.8 mm long crack was detected by the monitoring system. From the optical microscope measurement, crack length was found 18.2 mm, thus showing a relatively good agreement between actual crack tip position and the corresponding value obtained by the monitoring system.

Remarks

From the two studies discussed above, it is clear that, in order to measure the entire profile of BFS along the backface of the bonded region, the use of an array of strain sensors is a feasible solution.

It is also seen that, the use of an array of FBG sensors yields more accurate results compared to that of strain gages, presumably due to the fact that array of FBG is much more compact (Figure 11), allowing to place all the sensors along the centerline of the specimen, thus minimizing the effect of any possible misalignment. Besides, this array of FBG sensors require less wiring (basically only the fiber containing the array of sensors). Therefore, for BFS technique for SHM purposes, an array of FBG sensors is more suitable, provided that feasible solutions are adopted for compensating the effect of temperature variations.

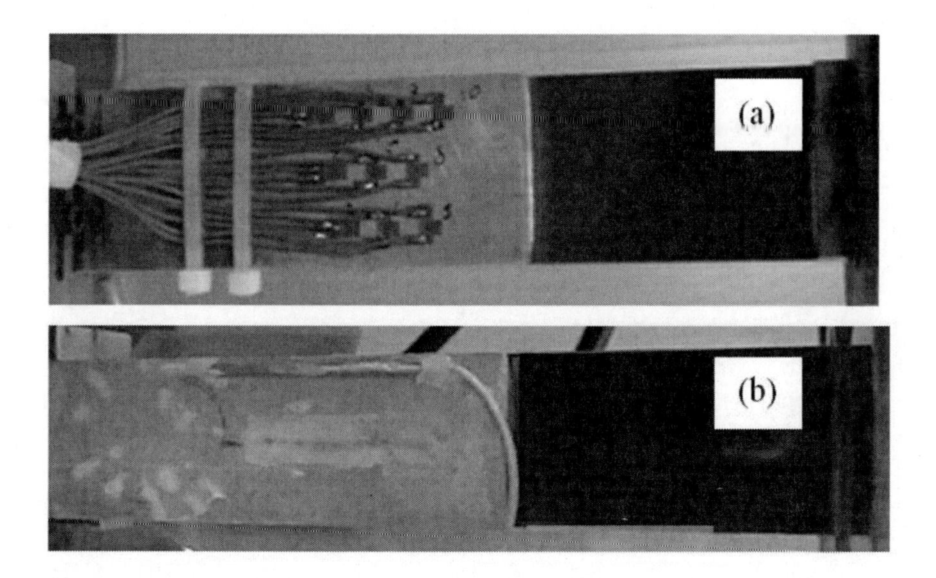

Figure 11. Visual comparison between (a) arrays of strain gauges and (b) an array of FBG sensors.

CONCLUSION

Adhesively bonded composite joints are used in various industrial applications. The integrity of the adhesive bond mainly depends on the quality of the interface between the adhesive and the substrate as well as the ability of the adhesive to sustain the applied loads during service life. Monitoring of structural integrity is of fundamental importance for such structures. By in-situ SHM methodologies, continuous real time monitoring might reduce maintenance costs while also increasing safety and reliability. Early detection of damage by using SHM techniques might also allow for extending the life of the structures, based on a damage tolerant design philosophy.

The BSF technique employing an array of sensors is one of the methods allowing for the detection of the initiation damage and continuous monitoring of fatigue cracks under service loads, at least for certain joint configurations, like the SLJ. Although both conventional electrical resistance strain gauges and optical FBG sensors can be used to construct array of strain sensors to detect BFS distribution for monitoring purposes, optical FBG offer many advantages in terms of compactness and ease of installation.

Acknowledgments

The authors gratefully acknowledge the contributions of Dr. Lorenzo Comolli, Dipartimento di Meccanica, Politecnico di Milano, to the development of the case studies discussed in this chapter.

References

[1] Tong L, Mouritz AP, Bannister MK. *3D Fiber Reinforced Polymer Composites.* ISBN 0-08-043938; Publisher: Elsevier Science Ltd, Oxford, UK. 2002. pp 1-12.

[2] Vasiliev VV, Morozov EV. *Mechanics and analysis of composite materials.* ISBN: 0-08-042702-2; Publisher: Elsevier Science Ltd, Oxford, UK. 2001. pp 9-26.

[3] Zangenberg J, Brøndsted P, Koefoed M. Mat. and Design. 2014, Vol 56, pp 635–641. Design of a fibrous composite preform for wind turbine rotor blades.

[4] Gianaris NJ. Automotive Applications, *Composites*, *ASM Handbook*, *ASM International*, Vol 21, 2001, pp 1020–1028.

[5] Mouritza AP, Gellertb E, Burchillb P, Challis K. Comp Struc July 2001, Vol 53, Issue 1, pp 21–42. Review of advanced composite structures for naval ships and submarines.

[6] Petrie EM. *Handbook of Adhesives and Sealants.* ISBN-10: 0071479163; Publisher: The McGraw-Hill Companies. 2007. Chapter 9: Surfaces and Surface preparation.

[7] Adams RD, Comyn J. Assembly Automation. 2000. Vol 20 Iss 2, 109 – 117. Joining using adhesives.

[8] Pethrick, RA. In *Comprehensive Composite Materials.* Editor, Kelly A, Zweben C. ISBN: 0-08 0429939. Publisher: Elsevier Science Ltd. 2000. Vol 5, pp 359-392.

[9] Hart-Smith LJ. *Analysis and Design of Advanced Composite Bonded Joints.* 1974, NASA CR-2218.

[10] Gao L, Chou TW, Thostenson ET, Zhang Z, Coulaud M. Carbon**.** Vol 49, Iss 10, Aug 2011, 3382–3385. In situ sensing of impact damage in epoxy/glass fiber composites using percolating carbon nanotube networks.

[11] Cai J, Qiu L, Yuan S, Shi L, Liu PP, Liang D. In *Composites and Their Applications*. Editor: Hu N. ISBN 978-953-51-0706-4. Publisher: InTech Open Access. 2012. Chap 3, pp 37-59.
[12] *Instrumentation Reference Book*. Editor: Boyes W. ISBN. 978-0-7506-8308-1. Publisher: Elsevier Inc. 2010. pp 567-592.
[13] Ducept F, Davies P, Gamby D. Int J Adhe Adhes, 2000, 20, 233-244. Mixed mode failure criteria for a glass/epoxy composite and an adhesively bonded composite/composite joint.
[14] Magalhaesa AG, de Moura MFSF. NDT&E Int. 2005, 38, 45–52. Application of acoustic emission to study creep behavior of composite bonded lap shear joints.
[15] Lim AS, Melrose ZR, Thostenson ET, Chou TW. Compos Sci Technol, 2011,71. 1183-1189. Damage sensing of adhesively-bonded hybrid composite/steel joints using carbon nanotubes.
[16] Baughman RH, Zakhidov AA, de Heer WA. Sci. 2002, 297, 787-792. Carbon Nanotubes-the Route Toward Applications.
[17] Thostenson ET, Chou TW. Adv Mater. 2006, 18, 2837-2841. Carbon Nanotube Networks: Sensing of Distributed Strain and Damage for Life Prediction and Self-Healing.
[18] Yu S, Tong MN, Critchlow G. J App Pol Sci. 2009, 111, 2957–2962. Wedge Test of Carbon-Nanotube-Reinforced Epoxy Adhesive Joints.
[19] Thostenson ET, Chou TW. Carbon. 2006, 44, 3022–3029. Processing-structure-multi-functional property relationship in carbon nanotube/epoxy composites.
[20] Thostenson ET, Chou TW. Nanotech, 2008, 19, 215713 (6pp). Real-time *in situ* sensing of damage evolution in advanced fiber composites using carbon nanotube networks.
[21] Nofar M, Hoa SV, Pugh MD. *Compos Sci Technol*. 2009, 69, 1599–1606. Failure detection and monitoring in polymer matrix composites subjected to static and dynamic loads using carbon nanotube networks.
[22] Montalvão D, Maia NMM, Ribeiro AMR. Shock Vib Dig. 2006, 38-4, July 295–324. A Review of Vibration-based Structural Health Monitoring with Special Emphasis on Composite Materials.
[23] Chiu WK, Galea SC, Koss LL, Rajic N. Smart. Mater Struct. 2000, 9, 466-475. Damage detection in bonded repairs using piezoceramics.
[24] Mickens T, Schulz M, Sundaresan M, Ghoshal A, Naser A, Reichmeider R. Mech Sys Signal Proc. 2003, 17, 285-303. Structural health monitoring of an aircraft joint.

[25] White C, Whittingham B, Li HC, Herszberg I, Mouritz AP. 2007. ACAM 2007, Eng Aus, 1, 198-203. Vibration based structural health monitoring of adhesively bonded composite scarf repairs.

[26] White C, Whittingham B, Li HC, Herszberg I, Mouritz AP. Comp. Struct.. 2009, 87, 175-181. Damage detection in repairs using frequency response techniques.

[27] Bernasconi A, Carboni M, Comolli L. Proc Eng. 2011, 10, 207–212. Monitoring of fatigue crack growth in composite adhesively bonded joints using Fiber Bragg Gratings.

[28] Bernasconi A, Comolli L. Proc. SPIE. 8421, 2012, 84214Y. An investigation of the crack propagation in a carbon-fiber bonded joint using backface strain measurements with FBG sensors.

[29] Haq M, Khomenko A, Udpa L, Udpa S. In *Experimental Mechanics of Composite, Hybrid, and Multifunctional Materials*. Editor: Tandon GP, Tekalur SA, Ralph C, Sottos NR, Blaiszik B. ISBN 978-3-319-00873-8. Publisher: Springer. Vol 6, Chap 22, pp 189-195.

[30] Herszberg I, Li HCH, Dharmawan F, Mouritz AP, Nguyen M, Bayandor *J. Comp. Struct.*, 2005, 67, 205–216. Damage assessment and monitoring of composite ship joints.

[31] Li HCH, Herszberg I, Davis CE, Mouritz AP, Galea SC. Comp. Struct., 2006, 75, 321-327. Health monitoring of marine composite structural joints using Fiber optic sensors.

[32] Silva-Munoz RA, Lopez-Anido RA. *Comp. Struct.*, 2009, 89, 224–234. Structural health monitoring of marine composite structural joints using embedded fiber Bragg grating strain sensors.

[33] Papantoniou A, Rigas G, Alexopoulos ND. Comp. Struct.. 2011, 93, 2163–2172. Assessment of the strain monitoring reliability of fiber Bragg grating sensor (FBGs) in advanced composite structures.

[34] Okabe Y, Mizutani T, Yashiro S, Takeda N. Compos Sci Technol. 2002, 62, 951–985. Detection of microscopic damages in composite laminate with embedded small-diameter fiber Bragg grating sensors.

[35] Takeda S, Okabe Y, Takeda N. *Compos: Part A.* 33, 2002, 971-980. Delamination detection in CFRP laminates with embedded small-diameter fiber Bragg grating sensors.

[36] Jones R, Galea S. *Comp. Struct.*, 2002, 58, 397-403. Health monitoring of composite repairs and joints using optical Fibers.

[37] Palaniappan J, Ogin SL, Thorne AM, Reed GT, Crocombe AD, Capell TF, Tjin SC, Mohanty L. *Compos Sci Technol.* 2008, 68, 2410-2417. Disbond growth detection in composite-composite single-lap joints using chirped FBG sensors.

[38] Schizas C, Stutz S, Botsis J, Coric D. *Comp. Struct.*, 2012, 94, 987-994. Monitoring of non-homogeneous strains in composites with embedded wavelength multiplexed fiber Bragg gratings: A methodological study.

[39] Zhang Z, Shang JK, Lawrence Jr. FV, *J of Adhes*, 1995, 49, 1-2. A Backface Strain Technique for Detecting Fatigue Crack Initiation in Adhesive Joints.

[40] Crocombe AD, Ong CY, Chan CM, Abdel Wahab MM, Ashcroft IA. J of Adhes, 2002, 78:9, 745-776. Investigating fatigue damage evolution in adhesively bonded structures using backface strain measurement.

[41] Shenoy V, Ashcroft IA, Critchlowb GW, Crocombe AD, Abdel Wahab MM. Int J Adhe Adhes, 2009, 29, 361-371. An investigation into the crack initiation and propagation behaviour of bonded single-lap joints using backface strain.

[42] Solana AG, Crocombe AD, Ashcroft IA. *Int J Adhe Adhes*, 2010, 30, 36-42. Fatigue life and backface strain predictions in adhesively bonded joints.

[43] Kreuzer M, Strain Measurement with Fiber Bragg Grating Sensors, HBM, Darmstadt , Germany.

[44] Othonos A. *Rev. Sci. Instrum*, 1997,68**,** 4309-4341. Fiber Bragg grating.

[45] Chan PKC, Jin W, Gong JM, Demokan MS. IEEE Pho Tech Lett, 1999, 1470-1472. Multiplexing of Fiber Bragg Grating Sensors Using an FMCW Technique.

[46] Bernasconi A, Kharshiduzzaman Md, Anodio, LF, Bordegoni M, Re GM, Braghin F, Comolli L. J of Adhe, 2013, Development of a Monitoring System for Crack Growth in Bonded Single-Lap Joints Based on the Strain Field and Visualization by Augmented Reality.

[47] Banea MD, da Silva LFM. Proc. ImechE, Part L, 223, 1-18. Adhesively bonded joints in composite materials: an overview.

[48] Comolli L, Micieli A. Proc. of Optical Fiber Sensing 2011, Ottawa, Canada.Numerical comparison of peak detection algorithms for the response of FBG in non-homogeneous strain fields.

[49] Peters K, Pattis P, Botsis J, Giaccari P. *Opt. Laser Eng*. 2000;33:107-19. Experimental verification of response of embedded optical fiber Bragg grating sensors in non homogeneous strain fields.

[50] Lomov SV, Ivanov DS, Verpoest I, Zako M, Kurashiki T, Nakai H, Molimard J, Vautrin A. *Composites: Part A* 39 (2008) 1218–1231. Full-field strain measurements for validation of meso-FE analysis of textile composites.

In: Adhesives
Editor: Dario Croccolo

ISBN: 978-1-63117-653-1

Chapter 6

THREE-DIMENSIONAL VS. TWO-DIMENSIONAL MODELLING OF FATIGUE DEBONDING OR DELAMINATION USING COHESIVE ZONE

A. Pirondi[1,*], G. Giuliese[1] and F. Moroni[2]
[1]Dipartimento di Ingegneria Industriale,
Università di Parma, Parma, Italy
[2]Centro Interdipartimentale SITEIA.PARMA,
Università di Parma, Parma, Italy

ABSTRACT

Adhesively bonded metallic and/or composite material structures are present as principal structural elements in several fields, such as aeronautics, automotive, nautical and wind energy. A numerical method able to reproduce three-dimensionally the fatigue debonding/delamination evolution in bonded structures is therefore necessary to improve their performances.

In this work the cohesive zone model previously developed by the authors to simulate fatigue crack growth at interfaces in 2D geometries is extended to 3D cracks under mixed-mode I/II loading.

[*] Corresponding Author address: Email:alessandro.pirondi@unipr.it.

Keywords: Fatigue, cohesive zone, finite element analysis

INTRODUCTION

Composite and hybrid metal/composite structures are nowadays present not only in the aerospace industry, but thanks to continuous performance improvement and cost reduction, also many more industrial fields are approaching the use of multimaterial structural elements. This requires, in turn, extensive use of adhesive bonding and a more and more sophisticated capability to simulate and predict the strength of bonded connections where, for this purpose, analytical methods are being progressively integrated or replaced by Finite Element Analysis (FEA). In engineering applications, it is well established that fatigue is the root cause of many structural failures. In the case of bonded joints, fatigue life is related to the initiation and propagation of defects starting at free edges of joining regions or other features, such as through-thickness holes. In the case of composite or metal/composite joints, fatigue can start also from defects at the same locations cited above, with the difference that the crack may either run into the adhesive or become a delamination crack. Especially in the case of damage tolerant or fail safe design, it is necessary to know how cracks, or in general defects, propagate during the service life of a component. A numerical method able to reproduce three-dimensionally the fatigue debonding in structures is therefore necessary to improve their performances.

The relationship between the applied stress intensity factor and the fatigue crack growth (FCG) rate of a defect is generally expressed as a power law [1]. In the case of polymers, adhesives and composites, the relationship is traditionally written as a function of the range of strain energy release rate (ΔG) as

$$\frac{da}{dN} = B\Delta G^{d} \tag{1}$$

where B and d are parameters depending on the material and load mixity ratio, and a is the defect length. In this simple form, the presence of a fatigue crack growth threshold and an upper limit to ΔG for fracture are not represented although, when needed, expressions accounting for these limits can be easily found (see for example [2]). In the same way, the influence of the stress ratio,

R, on the fatigue crack growth rate can be introduced into Eq.(1) by a term derived from extensions of the Paris law expressed in terms of the range of stress intensity factor, ΔK [3].

When a solution for the strain energy release rate as a function of crack length exists, then the number of cycles to failure comes out from the numerical integration between the initial crack length (a_0) and the final crack length (a_f) of the inverse of Eq.(1) ([2], [4]).

When a theoretical solution for the strain energy release rate does not exist, Finite Element (FE) simulation is commonly used to compute it. The prediction of crack growth can be then carried out by a stepwise analysis, each step corresponding to a user-defined crack growth increment and the number of cycles is obtained by integrating the crack growth rate computed from the Paris law. To speed up the process, in some finite element softwares, this procedure is integrated in special features (for example the *Debonding procedure in Abaqus®, Dassault Systèmes, Paris, France), where the strain energy release rate is obtained using the Virtual Crack Closure Technique (VCCT). An alternative way for dealing with fatigue crack growth problems is using the cohesive zone model (CZM). This model is commonly used for the simulation of the quasi static fracture problems, especially in the case of interface cracks such as in bonded joints and delamination in composites ([5]-[7] among others). The possibility to simulate the growth of a crack without any remeshing requirements and the relatively easy possibility to manipulate the constitutive law of the cohesive elements makes the cohesive zone model attractive also for the fatigue crack growth simulation [8]-[19]. However, differently from VCCT, three-dimensional fatigue debonding/delamination with CZM is not yet state-of-art in finite element softwares. Using [12] as a reference, but modifying the damage definition, including an automatic strain energy release rate evaluation and introducing different mixed mode criteria for the computation of the fatigue crack growth rate, the authors developed a model able to correctly predict fatigue crack growth at interfaces in two-dimensional geometries [19]. The extension of the model to full 3D cracks undergoing mixed-mode I/II fatigue loading is presented in this work, with a special emphasis on the changes done with respect to the 2D model.

MODEL FORMULATION

Although, different and complicated shapes of the cohesive law are proposed in the literature, the triangular one (Figure 1) is taken as it is often good enough to describe crack growth behaviour. In that case, damage starts once the tripping stress s_{ijmax} has been attained, decreasing progressively the element stiffness Kij.

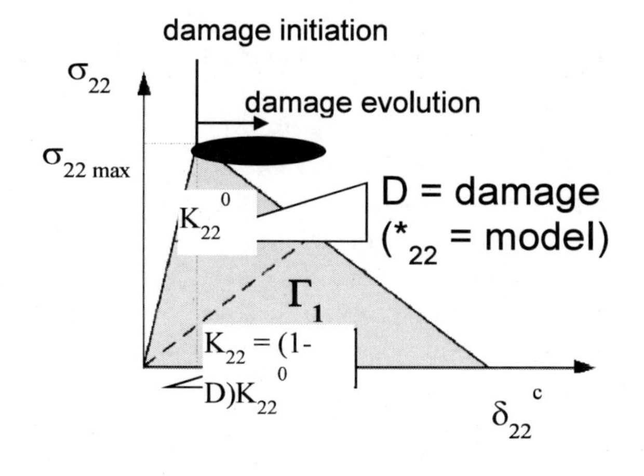

Figure 1. Example of a triangular cohesive law.

However, the model “as is” does not incorporate subcritical damage accumulation, i.e., cyclic damage is not possible below σ_{ij0} and does not accumulate beyond the first cycle above. In this model, the concept proposed in [12] is retained, while fundamental differences with respect to that work concern: i) the damage D is related directly to its effect on stiffness and not to the ratio between the energy dissipated during the damage process and the cohesive energy and then, in turn, to the stiffness; ii) the process zone size ACZ is defined as the sum of Ae of the cohesive elements for which the difference in opening between the maximum and minimum load of the fatigue cycle, $\Delta\delta = \delta_{max} - \delta_{min}$, is higher than a threshold value $\Delta\delta^{th}$; therefore, it is evaluated by FEA during the simulation and not derived from a theoretical model; iii) the strain energy release rate is calculated using the contour integral method over the cohesive process zone and it is implemented as a user-defined field subroutine (USDFLD) in Abaqus acting on standard cohesive elements, instead of a user element.

Considering a Representative Surface Element (represented in the simulation by a the cohesive element section pertaining to one integration point, Figure 2) with a nominal surface equal to A_e, the accumulated damage can be related to the damaged area due to micro voids or crack (A_d):

$$D = \frac{A_d}{A_e} \tag{2}$$

In [12], D is related to the ratio between the energy dissipated during the damage process and the cohesive energy (Γ_{22} in Figure 1) and then, in turn, to the stiffness. In the present model instead, D acts directly on stiffness, alike in [20].

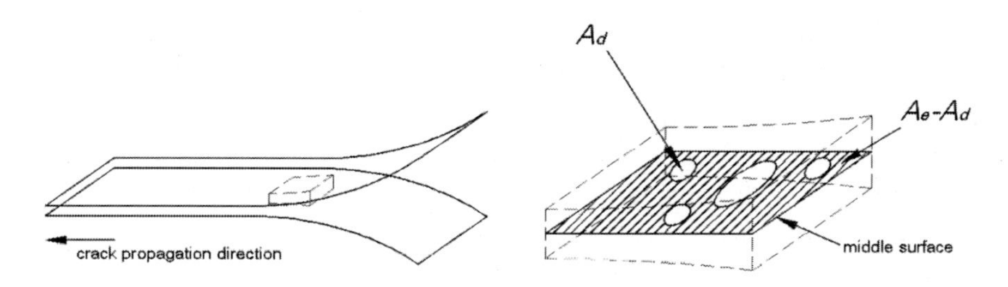

Figure 2. Nominal and damaged area in a representative surface element (RSE).

Referring to a mode I loading case, when the opening is relatively small the cohesive element behaves linearly; this happens until a given value of displacement, $\delta_{22,0}$ (or equivalently until a certain value of stress $\sigma_{22,0}$). This initial step is characterized by a stiffness $K_{22,0}$, that remains constant until $\delta_{22,0}$. Beyond this limit the stiffness is progressively reduced by D, until the final fracture in $\delta_{22,C}$ where the two surfaces are completely separated. Between $\delta_{22,0}$ and $\delta_{22,C}$ the stiffness K_{22} can be computed as

$$K_{22} = K_{22,0}(1 - D) \tag{3}$$

The area Γ_{22} underling the cohesive law is the energy to make the defect grow of a unit area and it is therefore representative of the fracture toughness, G_{IC}.

$$\Gamma_{22} = \int_0^{\delta_C} \sigma_{22} d\delta_{22} \tag{4}$$

In the monotonic case, the damage variable D can be written as a function of the opening (δ_{22}) and of the damage initiation and critical opening (respectively $\delta_{22,0}$ and $\delta_{22,C}$):

$$D = \frac{\delta_{22,c}\left(\delta_{22} - \delta_{22,0}\right)}{\delta_{22}\left(\delta_{22,c} - \delta_{22,0}\right)} \tag{5}$$

When the element is unloaded, the damage cannot be healed, therefore, looking at Figure 1, the unloading and subsequent loadings will follow the dashed line, until a further damage is attained. This simple model is able to describe the monotonic damage in case of mode I loading.

Considering the entire cohesive layer, the areal crack extension (A) can be computed as the sum of damaged areas of all the cohesive elements integration points (A_d) [12]

$$A = \sum A_d \tag{6}$$

When the fatigue damage is considered, from the previous equation, the crack growth (dA) can be written as a function of the increment of the damage area of all the cohesive elements (dA_d), therefore:

$$dA = \sum dA_d \tag{7}$$

However the damage increment would not concern the whole cohesive layer, but it will be concentrated in a relatively small process zone close to the crack tip. In order to estimate the size of A_{CZ}, analytical relationships can be found in the literature [21], where the size per unit thickness is defined as the distance from the crack tip to the point where $\sigma_{22,0}$ is attained. In this work, a different definition and evaluation method is proposed: A_{CZ} corresponds to the sum of the nominal sections of the cohesive elements where the difference in opening between the maximum and minimum load of the fatigue cycle, $\Delta\delta_{22} = \delta_{22,\max} - \delta_{22,\min}$, is higher than a threshold value $\Delta\delta_{22}^{th}$. The value is supposed to be the highest value of in the cohesive layer when ΔG in the simulation equals ΔG_{th} experimentally obtained by FCG tests. It has to be underlined that in this way FCG may take place even at $\delta_{22,\max} \le \delta_{22,0}$, which is a condition that should be accounted for since $\delta_{22,0}$ results from the calibration of cohesive zone on fracture tests and may not be representative of a threshold for FCG. The process zone size A_{CZ} has therefore to be evaluated by FEA while performing the FCG simulation but, on the other hand, does not need to be assumed from a theoretical model.

Eq.(7) can be therefore rewritten as [12]

$$dA = \sum_{i \in A_{CZ}} dA_d^{\,i} \tag{8}$$

where only the elements lying in the process zone (namely A_{CZ}) are considered.

In order to represent the crack growth due to fatigue (dA/dN), the local damage of the cohesive elements (D) has to be related to the number of cycles (N). This is done using thc equation

$$\frac{dD}{dN} = \frac{dD}{dA_d}\frac{dA_d}{dN} \tag{9}$$

The first part of Eq. (9) can be easily obtained deriving Eq. (2): therefore

$$\frac{dD}{dA_d} = \frac{1}{A_e} \tag{10}$$

The process to obtain the second part is quite more complicated: the derivative of Eq.(8) with respect to the number of cycles is

$$\frac{dA}{dN} = \sum_{i \in A_{CZ}} \frac{dA_d^i}{dN} \tag{11}$$

At this point an assumption is introduced: the increment of damage per cycle is supposed to be the same for all the elements lying in the process zone. Therefore the value dA_d/dN is assumed to be the average value of the damaged area growth rate dA_d^i/dN for all of the elements in the process zone.

Hence the crack growth rate can be rewritten as [12]:

$$\frac{dA}{dN} = \sum_{i \in A_{CZ}} \frac{dA_d}{dN} = n_{CZ} \frac{dA_d}{dN} \tag{12}$$

where n_{cz} is the number of elements lying on the process area A_{CZ}. n_{cz} can be written as the ratio between the process zone extension (A_{CZ}) and the nominal cross section area (A_e) leading to the equation

$$\frac{dA}{dN} = \frac{A_{CZ}}{A_e} \frac{dA_d}{dN} \tag{13}$$

The second part of Eq. (9) can be therefore written as:

$$\frac{dA_d}{dN} = \frac{dA}{dN} \frac{A_e}{A_{CZ}} \tag{14}$$

Combining Eqs. (10) and (14), the crack growth rate can be finally expressed as a function of the applied strain energy release rate, in the simplest version using Eq.(1)

$$\frac{dD}{dN} = \frac{1}{A_{CZ}} B \Delta G^d \tag{15}$$

STRAIN ENERGY RELEASE RATE COMPUTATION

The relationship between the applied strain energy release rate and the increase of damage in the cohesive zone needs a general method to calculate the value of the strain energy release rate as a function of crack length. The most common methods for the strain energy release rate evaluation by using the FEA are the contour integral (J) and the VCCT. These two methods are usually available in finite element softwares, but VCCT is intended in general as alternative to using cohesive elements while the software used in this work

(Abaqus® v. 6.11) does not output the contour integral for an integration path including cohesive element.

In order to compute the J-integral, a path surrounding the crack has to be selected.

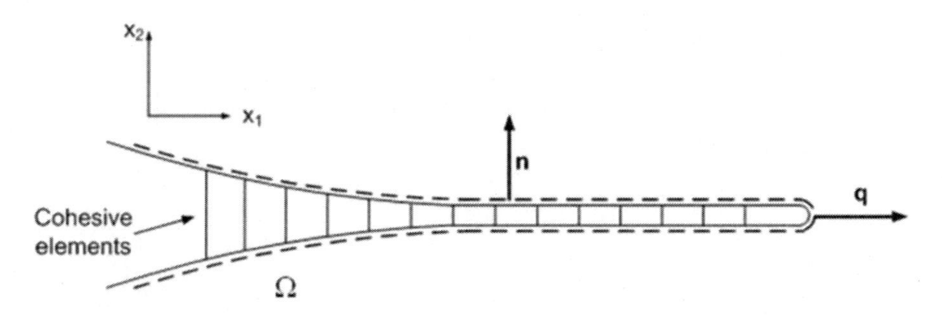

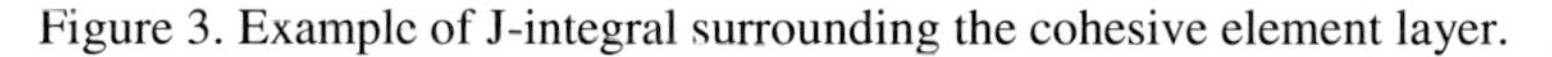
Figure 3. Example of J-integral surrounding the cohesive element layer.

Considering for example the crack in Figure 3, the path (Ω) is displayed by the dashed line and it is represented by the top and bottom nodes of the cohesive elements.

The J-integral definition [22] is

$$J = \int_{\Omega} \mathrm{n} \cdot [H] \cdot \mathrm{q} d\Omega \tag{16}$$

where n is a vector normal to the path, q is a vector lying on the crack propagation direction, and $[H]$ is defined as

$$[H] = W[I] - [\sigma_{ij}]\left[\frac{\partial u_{ij}}{\partial x_{ij}}\right] \tag{17}$$

where W is the strain energy density, the stress matrix and u_i the displacements of the points lying on the path.

Neglecting geometrical nonlinearity, the vector q can be assumed to be perpendicular to the direction x_2 along the whole path, therefore the J-integral can be rewritten as:

$$J = \int_{\Omega} \left(-\sigma_{12} \frac{\partial u_1}{\partial x_1} - \sigma_{22} \frac{\partial u_2}{\partial x_1} \right) d\Gamma \qquad (18)$$

Extracting the opening/sliding and the stresses in the cohesive elements at the beginning of the increment, the strain energy release rate is then computed. An interesting feature of this approach is that the mode I and the mode II component of the J-integral can be obtained by integrating separately the second or the first components of the integral in Eq.(18), respectively.

This method can be easily implemented for a two-dimensional problem, since there is only one possible path. In the case of three dimensional problems, the implementation is more difficult since several paths can be identified along the crack width, and moreover their definition is rather troublesome, especially when dealing with irregular meshes. In this work, a three-dimensional version has been implemented for the case of planar crack geometries and regular cohesive mesh, for which Eq.(18) is evaluated on several parallel contours in order to obtain J-integral along the crack front.

FINITE ELEMENT IMPLEMENTATION

The theoretical framework described in the first section and the strain energy release rate (SERR) calculation procedures are implemented programming Fortran subroutines templates available in the commercial software Abaqus®. In particular the USDFLD Abaqus® subroutine is used to modify the cohesive element stiffness by means of a field variable that accounts for damage, while the URDFIL subroutine is used to get the result in terms of stresses, displacements and energies. The fatigue analysis is carried out a as a simple static analysis, divided in a certain number of increments. Each increment corresponds to a given number of cycles.

Assuming that the fatigue cycle load varies from a maximum value P_{max} to a minimum value P_{min}, the analysis is carried out applying to the model the maximum load P_{max}. The load ratio is defined as the ratio between the minimum and maximum load applied

$$\Delta G = \left(1 - R^2\right) G_{\max} \qquad (19)$$

The strain energy release rate amplitude is therefore

$$R = \frac{P_{max}}{P_{min}} \tag{20}$$

This latter is compared with the strain energy release rate threshold ΔG_{th}. If $\Delta G > \Delta G_{th}$ the analysis starts (or it continues is the increment is not the first) otherwise the analysis is stopped.

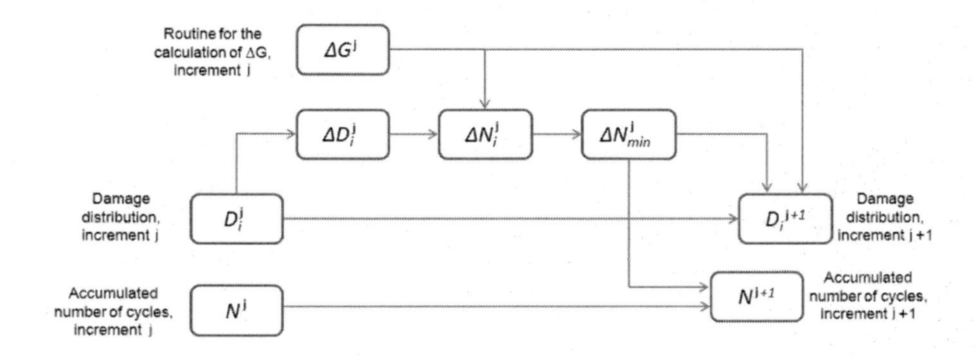

Figure 4. Flow diagram of the automatic procedure for the crack growth rate prediction.

The flow diagram in Figure 4 shows the operations done within each increment, where ΔD_i^j is calculated as follows:

$$\begin{aligned} \Delta D_i^j &= \Delta D_{\max} && if\ 1 - D_i^j > \Delta D_{\max} \\ \Delta D_i^j &= 1 - D_i^n && if\ 1 - D_i^j < \Delta D_{\max} \end{aligned} \tag{21}$$

In other words, ΔD_i^j is the minimum between the ΔD_{max} and the amount needed for D to reach the unity. The procedure is explained in detail in [19].

It is worth to underline that the procedure is fully automated, i.e., the simulation is performed in a unique run without stops. During the initial loading ramp, the stress-strain behavior is represented by the input cohesive law and the statically accumulated damage, represented by Eq. (5), is stored for the cohesive elements where $\delta > \delta_0$. The stiffness is then degraded according to the procedure described previously, starting from the stored static value until $D = 1$. If δ_{max} of the cycle is lower than δ_0 but $\Delta\delta > \Delta\delta_{th}$, elements entering the process zone during the fatigue analysis step start damaging before $\delta > \delta_0$. In this case the stress deviates from the linear elastic region of the cohesive law to fall on the linear damage part at $\delta < \delta_0$, i.e., without reaching σ_{max}.

Whenever a static overload occurs at a certain point in life, the cohesive element responds elastically with a degraded stiffness K instead of K_0, and further damage, or even static crack growth, can be accumulated related to the overload according to Eq. (5). After the overload, cyclic loading starts from the value of cyclic + static damage stored previously, see [23].

MIXED MODE LOADING

With the aim to extend the model to mixed-mode I/II conditions, a mixed mode cohesive law has to be defined. This is done according to the scheme shown in Figure 5 from the knowledge of the pure mode I and pure mode II cohesive laws (the index 22, refers to opening or mode I direction, index 12 refers to sliding or mode II direction)

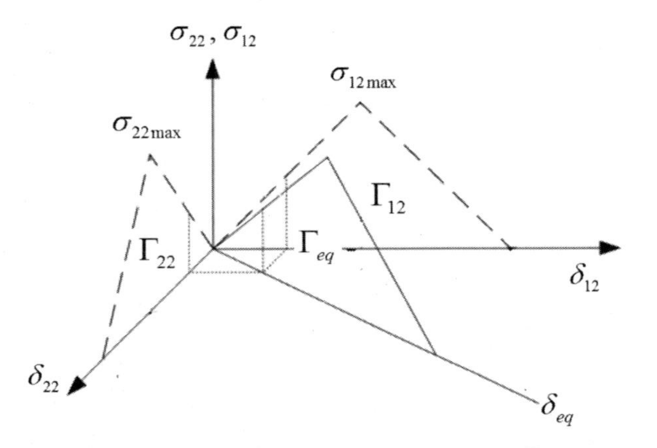

Figure 5. Example of cohesive law in the case of mixed mode conditions.

First of all the mixed mode equivalent opening has to be defined. This is done using the relationship

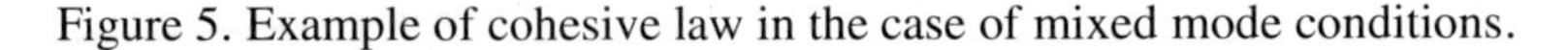

$$\delta_{eq} = \sqrt{\left(\frac{\delta_{22} + |\delta_{22}|}{2}\right)^2 + (\delta_{12})^2} \tag{22}$$

In case of pure mode I this equation yields the value of δ_{22} in case of positive δ_{22}, while it gives 0 in case of negative δ_{22}. This is done since it is supposed that compression stresses do not lead to the damage of the adhesive layer. Of course δ_{22} assumes only positive values if crack surface interpetration is properly prevented in the model. In the present model, interpenetration is prevented by restoring element stiffness if values of $\delta_{22} < 0$ are detected.

Moreover, the mixed mode cohesive law is defined in terms of the initial stiffness ($K_{eq,0}$), damage initiation equivalent opening ($\delta_{eq,0}$) and critical equivalent opening ($\delta_{eq,C}$). The equivalent initial stiffness is obtained by equating the equivalent strain energy (U_{EQ}) to the total strain energy (U_{TOT}), which in turn is equal to the sum of the strain energy in mode I (U_{22}) and in mode II (U_{12})

$$U_{EQ} = U_{TOT} = U_{22} + U_{12} = \frac{1}{2} \cdot \delta_{eq}^{\ 2} \cdot K_{eq}^{0} = \frac{1}{2} \cdot \left(\delta_{22} + |\delta_{22}|\right)^2 \cdot K_{22}^{0} + \frac{1}{2} \cdot \delta_{12}^{\ 2} \cdot K_{12}^{0} \tag{23}$$

$K_{22,0}$ and $K_{12,0}$ represent the initial stiffnesses of the mode I and mode II cohesive laws, respectively.

A further relationship is needed to define damage initiation: this is done using a the quadratic damage initiation criterion [24]

$$\left(\frac{\sigma_{22}}{\sigma_{22\max}}\right)^2 + \left(\frac{\sigma_{12}}{\sigma_{12\max}}\right)^2 = 1 \tag{24}$$

The last relationship needed, regards the definition of the critical equivalent opening. Since the area underlying the cohesive law is representative of the critical strain energy release rate, using the Kenane and Benzeggagh (KB) theory [25] the area underlying the mixed mode equivalent cohesive law (Γ_{eq}) can be computed as

$$\Gamma_{eq} = \Gamma_{22} + \left(\Gamma_{12} - \Gamma_{22}\right) \cdot MM^{m_m} \tag{25}$$

where (Γ_{22}) and (Γ_{12}) are the areas underlying the mode I and mode II cohesive laws, respectively, m_m is a mixed mode coefficient depending on the adhesive and MM is the mixed mode ratio defined as a function of the mode I and mode II strain energy release rates as follows:

$$MM = \frac{G_{II}}{G_I + G_{II}} \tag{26}$$

The KB mixed mode fatigue crack propagation model [26] is the first considered, since it is the most general law that can be found in the literature. The fatigue crack growth rate is given by Eq.(1) where this time B and d are functions of the mixed mode ratio MM:

$$\ln B = \ln B_{II} + (\ln B_I - \ln B_{II})(1 - MM)^{n_B} \tag{27}$$

$$d = d_I + (d_{II} - d_I)\cdot(MM)^{n_d} \tag{28}$$

d_I, B_I and d_{II}, B_{II} are, respectively, the parameters of the Paris law in mode I and mode II; $n_{d,}$ and n_B are material parameters. Also other literature approaches were implemented in [19] but they are not considered here for the sake of comparison with Abaqus® VCCT fatigue delamination, where KB is the only mixed mode loading FCG model. Moreover, updating of *B* and *d* with MM during propagation has been deactivated since it is not a feature available in Abaqus®.

FINITE ELEMENT MODELS

The CZM fatigue debonding model was tested on various joint geometries characterised by different mixed mode ratios, in order to verify accuracy, robustness and performance in terms of computational time. In particular, pure mode I loading was simulated with a Double Cantilever Beam (DCB) geometry, pure mode II loading with an End Loaded Split (ELS) geometry and mixed mode I/II loading with a Mixed Mode End Loaded Split (MMELS) geometry, as shown in Figure 6. The applied force and the specimens dimensions are given in Table 1. In order to investigate the model sensitivity to material behaviour, two kind of materials were simulated, one representing an elastic, isotropic aluminum alloy and another one an elastic, orthotropic composite laminate (see.Table 2). In all the simulations a load ratio $R = 0$ is assumed.

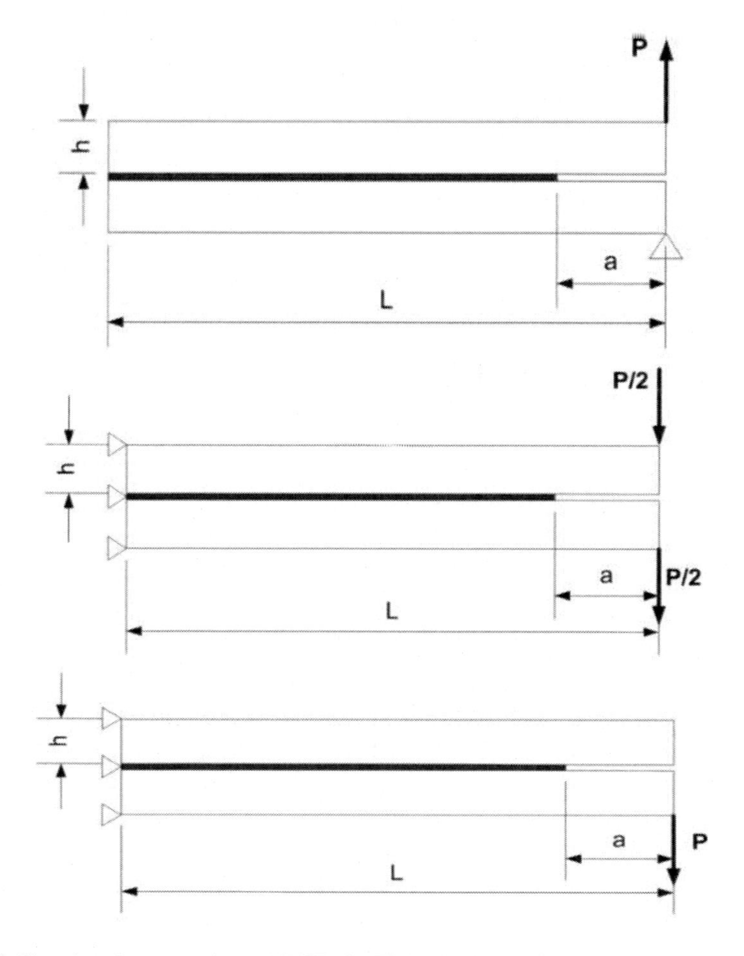

Figure 6. Simulated geometries: a) DCB, b) ELS, c) MMELS and d) SLJ.

Table 1. Specimen dimensions and applied load per unit thickness

	Aluminum Joints			Composite Joints		
	DCB	ELS	MMELS	DCB	ELS	MMELS
P [N/mm]	15	25	20	10	20	15
a_0 [mm]	20	20	20	20	20	20
h [mm]	5	5	5	5	5	5
L [mm]	175	175	175	175	175	175

The adherend were meshed with 3D Continuum Shell elements and the cohesive elements were kinematically tied to the two delaminating halves. The cohesive element size was 0.5 mm (CZM), while continuum shell elements were 1 mm in the direction of propagation, in order to keep computation time within a reasonable value. The maximum damage increment was taken ΔD_{max} = 0.2, based on the sensitivity analysis done in [28]. The increment in crack length per increment in analysis time came as a result of in the increment in damage ΔD, therefore it is not generally constant as ΔD may vary from increment to increment according to Eq. (26). However, the average increment in crack length ranged from 0.1 to 0.5mm in the various cases simulated in this work.

Table 2. Elastic constants, cohesive zone parameters and FCG behaviour for mode I, mode II, and mixed mode I/II [12], [27]

Debonding/Delamination			Aluminum		Composite	
Parameter	Mode I	Mode II	Parameter	Value	Parameter	Value
Γ [N/mm]	0.266	1.002	E [MPa]	70000	E_{11} [MPa]	54000
σ_{max} [MPa]	30	30	ν	0.29	E_{22} [MPa]	8000
δ_0 [mm]	0.003	0.003			ν_{12}	0.25
δ_C [mm]	0.0173	0.066			G_{12}[MPa]	2750
B	0.0616	4.23				
d	5.4	4.5	Parameter	Value		
			m_m	2.6		
			m_d	1.85		
			m_B	0.35		

RESULTS

The 3D model performance is compared with the 2D version [19] concerning the value of G as a function of crack length. Due to crack front bowing (Figure 7), the average G (or G_I, G_{II}) and crack length along the crack front were considered for the comparison with the 2D model. In the case of elastic, isotropic material parts, analytical solutions for G (Mode I [29], Mode II [30] and Mixed/Mode I/II [31]) were also introduced in the comparison, while in the case of the elastic, orthotropic composite the value of G obtained by quasi-static FE simulation using VCCT were considered.

Additionally, Eq.(1) is integrated numerically using the G vs. crack length coming from the analysis instead of taking directly the output number of cycles. The reason is that, as the CZM process zone needs some time to get to a steady state, the G calculated by CZM may be rather different from the analytical one in the first millimeters of propagation yielding a different number of cycles. For this reason, a comparison with experiments is foreseen as a further validation step, while at the moment the paper focuses on the comparison of numerical results of the present model and analytical solution after the transient phase of process zone formation.

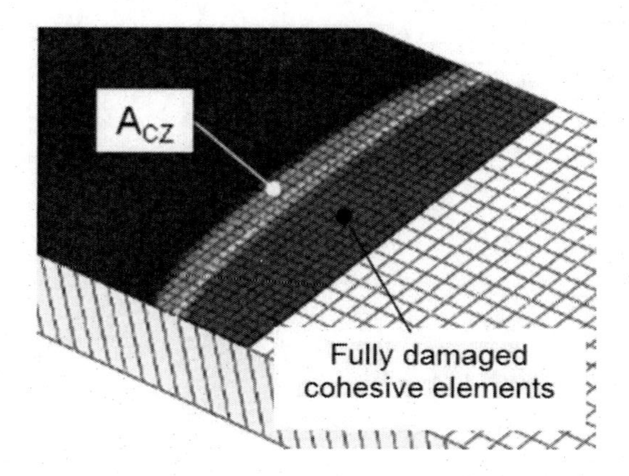

Figure 7. Bowed crack front and process zone in mode I fatigue modeling.

Mode I Loading (DCB)

In the case of aluminum joints, the four models show the same overall trend and an overall good correspondence with each other of the SERR plot, Figure 8 (a); in general the VCCT yields slightly lower values than the two CZM and analytical model. Moreover both the CZMs show a quite different trend in the first millimeter, where the process zone is under development. The average difference between the VCCT and the CZMs is about 5% on the 21-29 mm crack length span, while the difference is in the order of 2% whit respect to the analytic solution. As a consequence, the crack growth predictions given by the CZMs are rather similar to that given by the analytical model (Figure 8 (b)).

Due to the differences shown in the SERR plot, the VCCT model gives a slight under-prediction of the crack growth rate with respect to the CZMs and analytical model. It is believed that further mesh refinement can get the two models closer to each other.

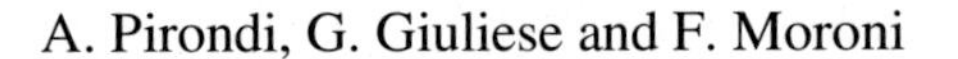

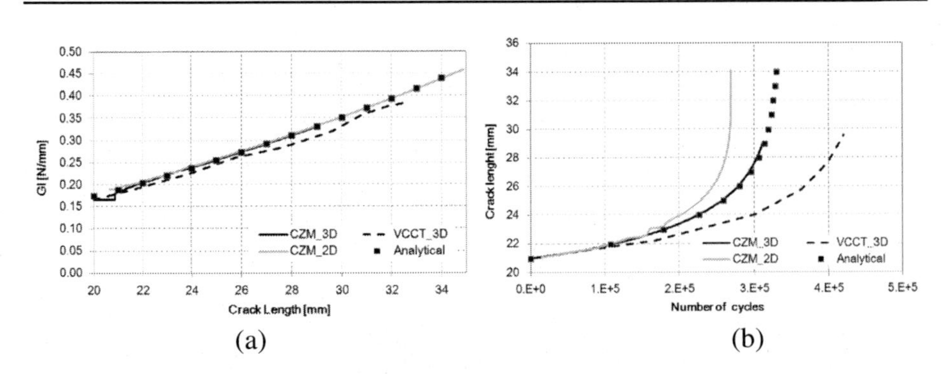

Figure 8. Comparison of (a) G_I and (b) da-dN obtained by CZM 2D, CZM 3D, VCCT 3D and analytical solution [29] in the case of DCB aluminum joints.

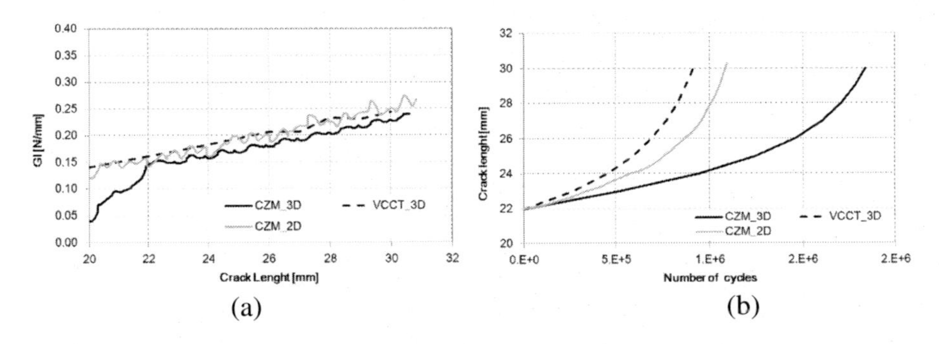

Figure 9. Comparison of (a) G_I and (b) da-dN obtained by CZM 2D, CZM 3D and VCCT 3D in the case of DCB composites joints.

Moving to the analysis of composite joints (Figure 9), the presence of an anisotropic behavior of the adherends increases the differences between the three models.

In particular, for the strain energy release rate evaluation, the trends of the CZMs are more scattered (Figure 9 (a)): this is due to the continuous changes produced in the stress field at the crack tip by the combination of adherends anisotropy and stepwise element deletion.

However a good agreement can be noticed for the three plots, although the CZM_3D curve is slightly lower than the other (differences in the order of 12%). These differences produce a significant under prediction of the crack growth rate in the case of CZM_3D with respect to CZM_2D and VCCT (Figure 9 (b)).

Mode II Loading (ELS)

Figure 10 (a) shows the values of G_{II} obtained by CZMs, VCCT and the analytical model in the case of aluminum substrates. Again as in the case of DCB, the four sets show a good overall correspondence with each other.

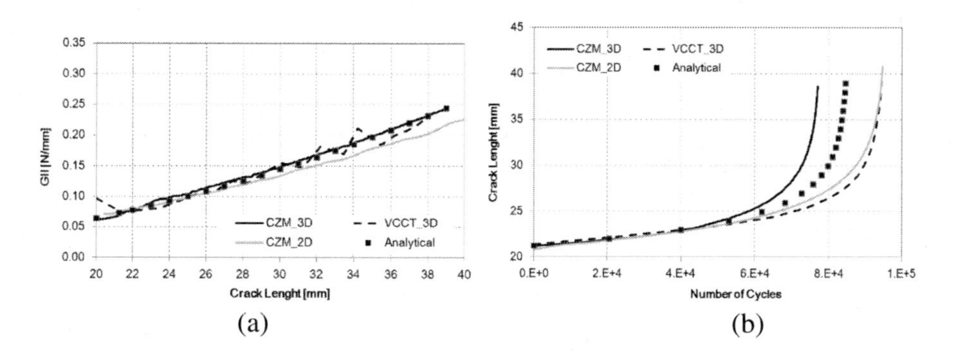

Figure 10. Comparison of (a) G_{II} and (b) da-dN obtained by CZM 2D, CZM 3D, VCCT 3D and analytical solution [30] in the case of ELS aluminum joints.

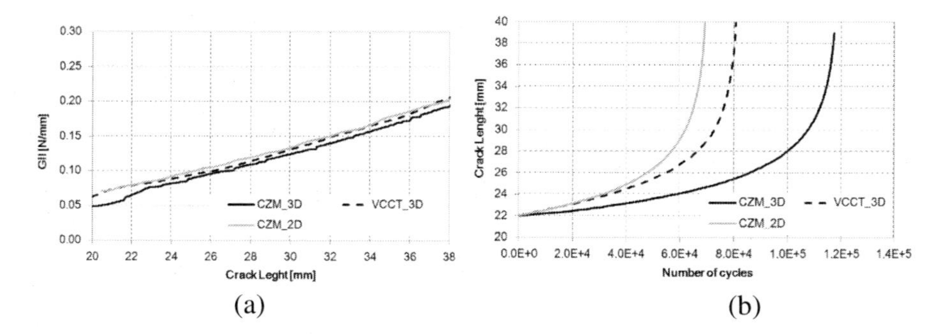

Figure 11. Comparison of (a) G_{II} and (b) da-dN obtained by CZM 2D, CZM 3D and VCCT 3D in the case of ELS composite joints.

Some differences concern the 2D CZM, that gives a trend slightly lower than the other models, and the VCCT, whose trend that is rather jagged: probably this is related to a non-uniform (one row of elements per time increment) crack front propagation that has been recorded in the simulation.

The results in terms of crack length vs. number of cycles are shown instead in Figure 10 (b), where differences come out lower than mode I (18% at 39 mm crack length); also in this case a further mesh refinement can get the four models closer to each other.

In the case of composite adherend the three models analyzed are again in good agreement between each other. Concerning the strain energy release rate plot (Figure 11 (a)), the CZM_3D trend is quite lower than the other; however the differences are, in the average, lower than 7%. In terms of crack length vs. number of cycles prediction (Figure 11 (b)), this produce similar results for CZM_2D and VCCT, while the CZM_3D yields a higher fatigue life with respect to the others (about 55% for a propagation from 22 to 36mm). The strain energy release rate plot of mode II composite joints appear less scattered with respect to the mode I composite joints: this is due to the lower stress concentration in the ELS joints (respect to the DCB), which produces smaller changes of the stress field when one or more elements change form "undamaged" to "damaged".

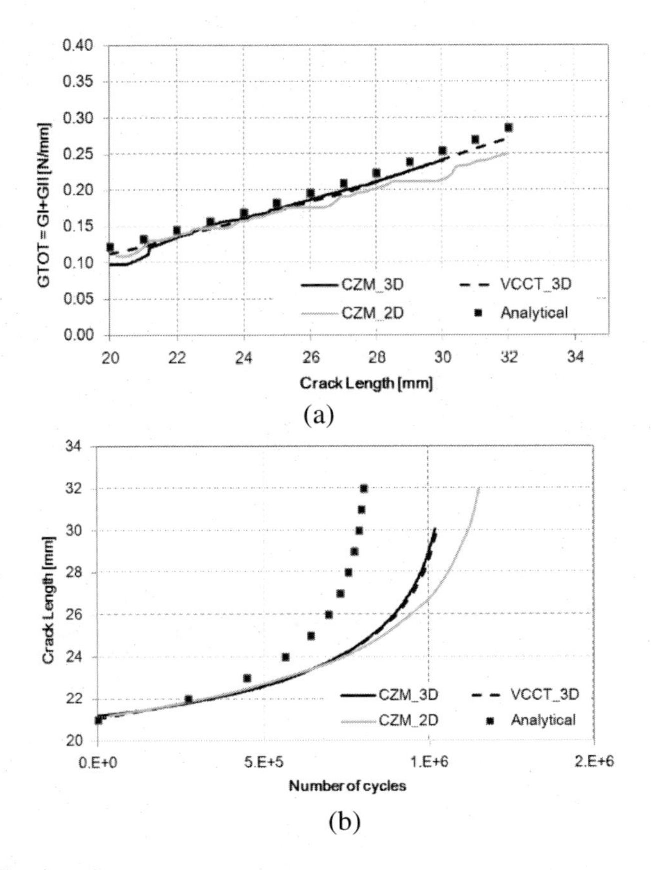

Figure 12. (Continued).

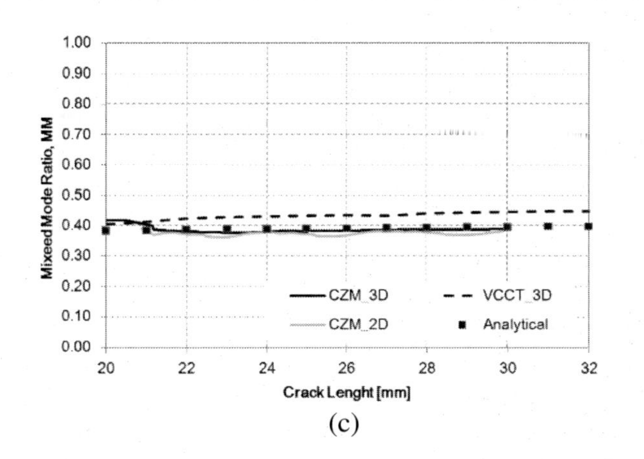

Figure 12. Comparison of (a) G_{II}, (b) da-dN and (c) mixed mode ratio obtained by CZM 2D, CZM 3D, VCCT 3D and analytical solution [31] in the case of MMELS aluminum joints.

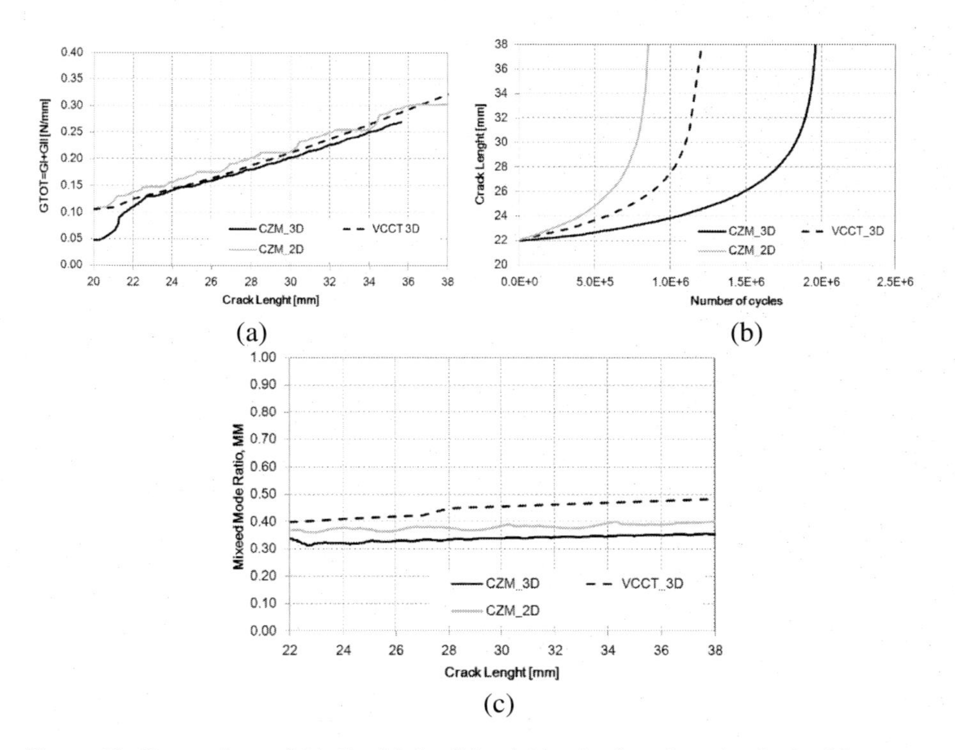

Figure 13. Comparison of (a) G_{II}, (b) da-dN and (c) mixed mode ratio obtained by CZM 2D, CZM 3D and VCCT 3D in the case of MMELS composite joints.

Mixed-Mode I/II Loading (MMELS)

Figure 12 (a) shows the values of G_{TOT} ($G_{TOT} = G_I + G_{II}$) obtained again by CZMs, VCCT and analytical model for the aluminum joint.

The values obtained with the four methods in this case highlight a little bit the differences found for the single modes: the 3D CZM trend is almost superimposed to the VCCT trend, while the analytical solution gives a slight overestimation and the 2D CZM a slight underestimation with respect to the previous ones.

These differences become larger in the crack length vs. number of cycles plot (Figure 12 (b)), where again CZM 3D and VCCT give similar result, while analytical solution and the 2D CZM give respectively higher and lower crack growth rate. Figure 12 (c) shows the variation of the mixed mode ratio (MM) during the propagation. For the four models, the trend is quite similar and almost constant close of a value of 0.4. Only in the VCCT model the MM value results quite higher than the other (the differences in the order of 9%). In the case of composite joints (Figure 13) the three models give quite different results: although the trends of the total strain energy release rate ($G_{TOT}=G_I+G_{II}$) are similar (Figure 13 (a)), the mode separation produces differences of the mixed mode ratio in the order of 30% (Figure 13 (c)), and this, in turn, produces significant differences in terms of crack length vs. number of cycles prediction (Figure 13 (b)). This phenomenon, still under investigation, is thought to be produced by the material anisotropy.

CONCLUSION

The performances of the 3D CZM model were compared with a previously developed 2D model and analytical solutions on mode I, mode II and mixed-mode I/II loaded cracks in bonded aluminum or composite assemblies. The results are in an overall good agreement with each other, with the exception of the first instants of propagation of the 3D model, where the CZM process zone has to shape up. Next step should to assess this initial transient of 3D process zone formation, leading to a higher predicted number of cycles with respect to 2D simulation, by experimental evidence.

ACKNOWLEDGMENTS

The work of F. Moroni was partially supported by Emilia-Romagna Region (POR FSE 2007-2013).

REFERENCES

[1] Paris P, Erdogan F (1961), A critical analysis of crack propagation laws, *J. Basic Eng.,* 85, 528-534.

[2] Curley A J, Hadavinia A, Kinloch A J, Taylor A C (2000), Predicting the service-life of adhesively-bonded joints, *International Journal of Fracture*, 103, 41-69.

[3] Forman F G, Kearnay V E, Engle R M (1967), Numerical analysis of crack propagation in cyclic loaded structures, *J. Basic Eng.*, Trans ASME 89, 885.

[4] Pirondi A, Moroni F (2009), An investigation of fatigue failure prediction of adhesively bonded metal/metal joints, *International Journal of Adhesion & Adhesives*, 29, 796–805.

[5] Hutchinson J W, Evans A G (2000), Mechanics of materials: top-down approaches to fracture, *Acta Materialia*, 48, 125-135.

[6] Blackman B R K, Hadavinia H, Kinloch A J, Williams J G (2003), The use of a cohesive zone model to study the fracture of fibre composites and adhesively-bonded joints, *International Journal of Fracture*, 119, 25-46.

[7] Li S, Thouless M D, Waas A M, Schroeder J A and Zavattieri P D (2005), Use of Mode-I cohesive-zone models to describe the fracture of an adhesively-bonded polymer-matrix composite, *Composite Science and Technology*, 65, 281-293.

[8] Maiti S, Geubelle P H (2005), A cohesive model for fatigue failure of polymers. *Engineering Fracture Mechanics*, 72, 691-708.

[9] Yang Q D, Thouless M D, Ward S M (1999), Numerical Simulations of Adhesively-Bonded Beams Failing with Extensive Plastic Deformation, *Journal of Mechanics and Physics of Solids*, 47, 1337-1353.

[10] Roe K L, Siegmund T (2003), An irreversible cohesive zone model for interface fatigue crack growth simulation, *Engineering Fracture Mechanics*, 70, 209-232.

[11] Muñoz J J, Galvanetto U, Robinson P (2006), On the numerical simulation of fatigue driven delamination with interface element, *International Journal of Fatigue*, 28, 1136-1146.

[12] Turon A, Costa J, Camanho P P, Dávila C G (2007), Simulation of delamination in composites under high-cycle fatigue, *Composites: Part A*, 38, 2270-2282.

[13] Khoramishad H, Crocombe A D, Katnam K B, Ashcroft I A (2010), Predicting fatigue damage in adhesively bonded joints using a cohesive zone model, *International Journal of Fatigue*,32, 1146–1158.

[14] Khoramishad H, Crocombe A D, Katnam K B, Ashcroft I A (2010), Fatigue damage modelling of adhesively bonded joints under variable amplitude loading using a cohesive zone model, *Engineering Fracture Mechanics*, 78, 3212-3225.

[15] Naghipour P, Bartsch M, Voggenreiter H (2011) Simulation and experimental validation of mixed mode delamination in multidirectional CF/PEEK laminates under fatigue loading, *International Journal of Solids and Structures*, 48, 1070–1081.

[16] P.W. Harper, S.R. Hallett (2010) A fatigue degradation law for cohesive interface elements – Development and application to composite materials, *International Journal of Fatigue,* 32, 1774–1787.

[17] P. Beaurepaire, G.I. Schuëller (2011) Modeling of the variability of fatigue crack growth using cohesive zone elements. *Engineering Fracture Mechanics*, 78, 2399–2413.

[18] G. Allegri, M.R. Wisnom (2012), A non-linear damage evolution model for mode II fatigue delamination onset and growth, *International Journal of Fatigue*, 43, 226–234.

[19] Moroni F, Pirondi A (2012), A procedure for the simulation of fatigue crack growth in adhesively bonded joints based on a cohesive zone model and various mixed-mode propagation criteria, *Engineering Fracture Mechanics*, 89, 129–138.

[20] Lemaitre J (1985), Continuous Damage Mechanics Model for Ductile Fracture. *Journal of Engineering Materials and Technology*, 107, 83-89.

[21] Harper W P, Hallett S R (2008), Cohesive zone length in numerical simulations of composite delamination. *Engineering Fracture Mechanics*, 75, 4774–4792.

[22] Rice J R (1968), A Path Independent Integral and the Approximate Analysis of Strain Concentration by Notches and Cracks, *Journal of Applied Mechanics,* 35, 379-386.

[23] Pirondi A., Moroni F. (2011) Simulating fatigue failure in bonded composite joints using a modified cohesive zone model. In Composite Joints and Connections, PP Camanho and L Tong Eds. Woodhead Publishing.

[24] Ungsuwarungsru T. and Knauss, W. G. (1987) The role of Damage-Softened Material Behaviour in the Fracture of Composites and Adhesives. *International Journal of Fracture* 35, pp. 221-241.

[25] Kenane M, Benzeggagh M L (1996), Measurement of mixed mode delamination fracture toughness of unidirectional glass/epoxy composites with mixed mode bending apparatus, *Composite Science and Technology*, 56, 439-449.

[26] Kenane M, Benzeggagh M L (1997), Mixed mode delamination fracture toughness of unidirectional glass/epoxy composites under fatigue loading. *Composite Science and Technology*, 57, 597-605.

[27] Bernasconi A, Jamil A, Moroni F, Pirondi A, (2013), A study on fatigue crack propagation in thick composite adhesively bonded joints, *International Journal of Fatigue*, 50, 18-25.

[28] Moroni F, Pirondi A (2010), A Progressive Damage Model for the Prediction of Fatigue Crack Growth in Bonded Joints, *The Journal of Adhesion*, 86, 1-21.

[29] Krenk S (1992), *Eng. Fract. Mech.,* 43, 549.

[30] Wang Y, Williams J.G. (1992), *Composites Sci. Technol.*, 34, 251

[31] Blanco N, Gamstedt E K, Costa J, Trias D Analysis of the mixed-mode end load split delamination test (2006) *Composite Structure* 76, 14-20.

INDEX

D

E

F

G

H

I

K

L

M

N

O

T

U

V

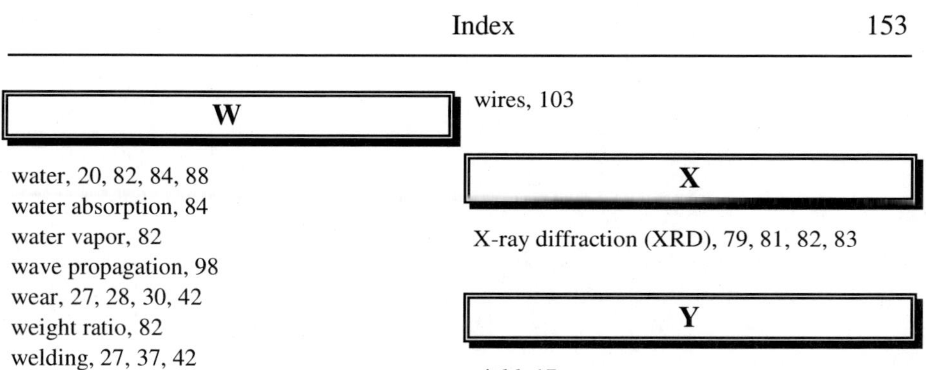

W

X

Y